IZC®

A Member of the International Code Family®

INTERNATIONAL **ZONING** CODE®

2018 International Zoning Code®

First Printing: August 2017

ISBN: 978-1-60983-755-6 (soft-cover edition)

Date of First Publication: August 31, 2017

T024800

PRINTED IN THE USA

PREFACE

Introduction

The *International Zoning Code*® (IZC®) establishes minimum requirements for zoning ordinances. This 2018 edition is fully compatible with all of the *International Codes*® (I-Codes®) published by the International Code Council (ICC®), including the *International Building Code*®, *International Energy Conservation Code*®, *International Existing Building Code*®, *International Fire Code*®, *International Fuel Gas Code*®, *International Green Construction Code*®, *International Mechanical Code*®, *International Plumbing Code*®, *International Private Sewage Disposal Code*®, *International Property Maintenance Code*®, *International Residential Code*®, *International Swimming Pool and Spa Code*®, *International Wildland-Urban Interface Code*® *and International Code Council Performance Code*®.

The I-Codes, including this *International Zoning Code*, are used in a variety of ways in both the public and private sectors. Most industry professionals are familiar with the I-Codes as the basis of laws and regulations in communities across the U.S. and in other countries. However, the impact of the codes extends well beyond the regulatory arena, as they are used in a variety of nonregulatory settings, including:

- Voluntary compliance programs such as those promoting sustainability, energy efficiency and disaster resistance.
- The insurance industry, to estimate and manage risk, and as a tool in underwriting and rate decisions.
- Certification and credentialing of individuals involved in the fields of building design, construction and safety.
- Certification of building and construction-related products.
- U.S. federal agencies, to guide construction in an array of government-owned properties.
- Facilities management.
- "Best practices" benchmarks for designers and builders, including those who are engaged in projects in jurisdictions that do not have a formal regulatory system or a governmental enforcement mechanism.
- College, university and professional school textbooks and curricula.
- Reference works related to building design and construction.

In addition to the codes themselves, the code development process brings together building professionals on a regular basis. It provides an international forum for discussion and deliberation about building design, construction methods, safety, performance requirements, technological advances and innovative products.

Development

This 2018 edition presents the code as originally issued, with changes reflected in the 2003 through 2015 editions and further changes approved by the ICC Code Development Process through 2016. A new edition of the code is promulgated every 3 years.

This code is intended to establish provisions that adequately protect public health, safety and welfare; that do not unnecessarily increase construction costs; that do not restrict the use of new materials, products or methods of construction; and that do not give preferential treatment to particular types or classes of materials, products or methods of construction.

Maintenance

The *International Zoning Code* is kept up to date through the review of proposed changes submitted by code enforcement officials, industry representatives, design professionals and other interested parties. Proposed changes are carefully considered through an open code development process in which all interested and affected parties may participate.

The ICC Code Development Process reflects principles of openness, transparency, balance, due process and consensus, the principles embodied in OMB Circular A-119, which governs the federal government's use of private-sector standards. The ICC process is open to anyone; there is no cost to participate, and people can participate without travel cost through the ICC's cloud-based app, cdpAccess®. A broad cross section of interests are represented in the ICC Code Development Process. The codes, which are updated regularly, include safeguards that allow for emergency action when required for health and safety reasons.

In order to ensure that organizations with a direct and material interest in the codes have a voice in the process, the ICC has developed partnerships with key industry segments that support the ICC's important public safety mission. Some code development committee members were nominated by the following industry partners and approved by the ICC Board:

- American Institute of Architects (AIA)
- National Association of Home Builders (NAHB)

The code development committees evaluate and make recommendations regarding proposed changes to the codes. Their recommendations are then subject to public comment and council-wide votes. The ICC's governmental members—public safety officials who have no financial or business interest in the outcome—cast the final votes on proposed changes.

The contents of this work are subject to change through the code development cycles and by any governmental entity that enacts the code into law. For more information regarding the code development process, contact the Codes and Standards Development Department of the International Code Council.

While the I-Code development procedure is thorough and comprehensive, the ICC, its members and those participating in the development of the codes disclaim any liability resulting from the publication or use of the I-Codes, or from compliance or noncompliance with their provisions. The ICC does not have the power or authority to police or enforce compliance with the contents of this code.

Code Development Committee Responsibilities (Letter Designations in Front of Section Numbers or Definitions)

In each code development cycle, proposed changes to the code are considered at the Committee Action Hearings by the International Property Maintenance/Zoning Code Development Committee, whose action constitutes a recommendation to the voting membership for final action on the proposed change. Proposed changes to a code section that has a number beginning with a letter in brackets are considered by a different code development committee. For example, proposed changes to definitions that have [BG] in front of them (e.g., [BG] DWELLING UNIT) are considered by the IBC—General Code Development Committee at the code development hearings.

The content of sections or definitions in this code that begin with a letter designation is maintained by another code development committee in accordance with the following:

[A] = Administrative Code Development Committee; and

[BG] = IBC—General Code Development Committee.

For the development of the 2021 edition of the I-Codes, there will be two groups of code development committees and they will meet in separate years.

Group A Codes (Heard in 2018, Code Change Proposals Deadline: January 8, 2018)	Group B Codes (Heard in 2019, Code Change Proposals Deadline: January 7, 2019)
International Building Code – Egress (Chapters 10, 11, Appendix E) – Fire Safety (Chapters 7, 8, 9, 14, 26) – General (Chapters 2–6, 12, 27–33, Appendices A, B, C, D, K, N)	Administrative Provisions (Chapter 1 of all codes except IECC, IRC and IgCC, administrative updates to currently referenced standards, and designated definitions)
International Fire Code	**International Building Code** – Structural (Chapters 15–25, Appendices F, G, H, I, J, L, M)
International Fuel Gas Code	**International Existing Building Code**
International Mechanical Code	**International Energy Conservation Code—Commercial**
International Plumbing Code	**International Energy Conservation Code—Residential** – IECC—Residential – IRC—Energy (Chapter 11)
International Property Maintenance Code	**International Green Construction Code** (Chapter 1)
International Private Sewage Disposal Code	**International Residential Code** – IRC—Building (Chapters 1–10, Appendices E, F, H, J, K, L, M, O, Q, R, S, T)
International Residential Code – IRC—Mechanical (Chapters 12–23) – IRC—Plumbing (Chapters 25–33, Appendices G, I, N, P)	
International Swimming Pool and Spa Code	
International Wildland-Urban Interface Code	
International Zoning Code	
Note: Proposed changes to the ICC *Performance Code*™ will be heard by the code development committee noted in brackets [ ] in the text of the ICC *Performance Code*™.	

Code change proposals submitted for code sections that have a letter designation in front of them will be heard by the respective committee responsible for such code sections. Because different committees hold Committee Action Hearings in different years, proposals for the IZC will be heard by committees in both the 2018 (Group A) and the 2019 (Group B) code development cycles.

For example, the definition of "Dwelling unit" in Section 202 is designated as the responsibility of the IBC—General Code Development Committee. As noted in the preceding table, that committee will hold its Committee Action Hearings in 2018 to consider code change proposals for the chapters for which it is responsible. Therefore, any proposals to this definition in Chapter 2 will need to be submitted by January 8, 2018, for consideration in 2018 by the appropriate International Building Code Committee (IBC—General).

As another example, every section of Chapter 1 of this code is designated as the responsibility of the Administrative Code Development Committee, which is part of the Group B portion of the hearings. This committee will hold its Committee Action Hearings in 2019 to consider code change proposals for Chapter 1 of all I-Codes except the *International Energy Conservation Code, International Residential Code* and *International Green Construction Code*. Therefore, any proposals received for Chapter 1 of this code will be assigned to the Administrative Code Development Committee for consideration in 2019.

It is very important that anyone submitting code change proposals understands which code development committee is responsible for the section of the code that is the subject of the code change proposal. For further information on the code development committee responsibilities, please visit the ICC website at www.iccsafe.org/scoping.

Marginal Markings

Solid vertical lines in the margins within the body of the code indicate a technical change from the requirements of the 2015 edition. Deletion indicators in the form of an arrow (➡) are provided in the margin where an entire section, paragraph, exception or table has been deleted or an item in a list of items or a table has been deleted.

Chapter 1, Scope and Administration, had no technical changes from the 2015 edition; however, the chapter has been reordered for consistency with the other I-Codes. The following table indicates the reordering of Chapter 1 of the 2018 edition of the *International Zoning Code*.

2018 ORDER	2015 ORDER
102	103
103	105
105	106
106	107
107	108
108	109
109	110
110	111
111	102

Coordination of the International Codes

The coordination of technical provisions is one of the strengths of the ICC family of model codes. The codes can be used as a complete set of complementary documents, which will provide users with full integration and coordination of technical provisions. Individual codes can also be used in subsets or as stand-alone documents. To make sure that each individual code is as complete as possible, some technical provisions that are relevant to more than one subject area are duplicated in some of the model codes. This allows users maximum flexibility in their application of the I-Codes.

Italicized Terms

Words and terms defined in Chapter 2, Definitions, are italicized where they appear in code text and the Chapter 2 definition applies. Where such words and terms are not italicized, common-use definitions apply. The words and terms selected have code-specific definitions that the user should read carefully to facilitate better understanding of the code.

Adoption

The International Code Council maintains a copyright in all of its codes and standards. Maintaining copyright allows the ICC to fund its mission through sales of books, in both print and electronic formats. The ICC welcomes adoption of its codes by jurisdictions that recognize and acknowledge the ICC's copyright in the code, and further acknowledge the substantial shared value of the public/private partnership for code development between jurisdictions and the ICC.

The ICC also recognizes the need for jurisdictions to make laws available to the public. All I-Codes and I-Standards, along with the laws of many jurisdictions, are available for free in a nondownloadable form on the ICC's website. Jurisdictions should contact the ICC at adoptions@iccsafe.org to learn how to adopt and distribute laws based on the *International Zoning Code* in a manner that provides necessary access, while maintaining the ICC's copyright.

To facilitate adoption, several sections of this code contain blanks for fill-in information that needs to be supplied by the adopting jurisdiction as part of the adoption legislation. For this code, please see:

Section 101.1. Insert [NAME OF JURISDICTION]

Section 108.2.2. Insert [NUMBER OF WORKING DAYS]

Table 302.1. Insert [MINIMUM AREAS]

Section 1008.1.1. Insert [SIGN AREA]

Table 1008.1.1(1). Insert [SIGN AREAS IN THREE LOCATIONS]

Table 1008.1.1(2). Insert [PERCENTAGE OF BUILDING ELEVATION IN THREE LOCATIONS]

Table 1008.1.2. Insert [NO. OF SIGNS, HEIGHT AND AREA IN 10 LOCATIONS]

Section 1008.1.3. Insert [SIGN AREAS IN TWO LOCATIONS]

Section 1008.2.1. Insert [SIGN AREAS IN EIGHT LOCATIONS]

Section 1008.2.2. Insert [SIGN HEIGHTS AND AREA IN 10 LOCATIONS]

Section 1008.2.3. Insert [SIGN AREAS IN THREE LOCATIONS]

Section 1008.2.5. Insert [SIGN HEIGHT AND AREA IN TWO LOCATIONS]

Section 1008.2.6. Insert [SIGN HEIGHT AND AREA IN TWO LOCATIONS]

Section 1008.3.3. Insert [SIGN AREA, HEIGHT, PROJECTION AND VERTICAL DISTANCE IN SIX LOCATIONS]

Section 1008.3.4. Insert [SIGN AREA AND VERTICAL DISTANCE IN TWO LOCATIONS]

Section 1008.3.5. Insert [SIGN HEIGHT IN TWO LOCATIONS]

EFFECTIVE USE OF THE INTERNATIONAL ZONING CODE

Effective Use of the International Zoning Code

The *International Zoning Code* (IZC) is a model code that regulates minimum zoning requirements for new buildings.

The IZC is a planning and community development document. The IZC is intended to provide for the arrangement of compatible buildings and land uses and establish provisions for the location of all types of uses, in the interest of the social and economic welfare of the community.

Arrangement and Format of the 2018 IZC

Before applying the requirements of the IZC, it is beneficial to understand its arrangement and format. The IZC, like other codes published by ICC, is arranged and organized to follow sequential steps that generally occur during a plan review or inspection. The IZC is divided into 14 different parts:

Chapter	Subject
1	Scope and Administration
2	Definitions
3	Use Districts
4	Agricultural Zones
5	Residential Zones
6	Commercial and Commercial/Residential Zones
7	Factory/Industrial Zones
8	General Provisions
9	Special Regulations
10	Sign Regulations
11	Nonconforming Structures and Uses
12	Conditional Uses
13	Planned Unit Development
14	Referenced Standards

The following is a chapter-by-chapter synopsis of the scope and intent of the provisions of the *International Zoning Code*:

Chapter 1 Scope and Administration. This chapter contains provisions for the application, enforcement and administration of subsequent requirements of the code. In addition to establishing the scope of the code, Chapter 1 identifies which buildings and structures come under its purview. Chapter 1 is largely concerned with maintaining "due process of law" in enforcing the zoning criteria contained in the body of the code. Only through careful observation of the administrative provisions can the building official reasonably expect to demonstrate that "equal protection under the law" has been provided.

Chapter 2 Definitions. Terms that are defined in the code are listed alphabetically in Chapter 2. While a defined term may be used in one chapter or another, the meaning provided in Chapter 2 is applicable throughout the code.

Additional definitions regarding signs are found in Chapter 10. These are not listed in Chapter 2.

Where understanding of a term's definition is especially key to or necessary for the understanding of a particular code provision, the term is shown in *italics* wherever it appears in the code. This is true only for those terms that have a meaning that is unique to the code. In other words, the generally understood meaning of a term or phrase might not be sufficient or consistent with the meaning prescribed by the code; therefore, it is essential that the code-defined meaning be known.

Guidance regarding tense, gender and plurality of defined terms as well as guidance regarding terms not defined in this code is provided.

Chapter 3 Use Districts. Chapter 3 identifies classifications for typical zoning districts and provides for the application of minimum district areas, in order to regulate and restrict the locations for uses and locations of buildings designated for specific areas and to regulate the minimum required areas or yards and courts and important open-areas property.

This chapter also requires coordination of the established zoning districts with approved zoning maps. Further, this chapter also contains information on the minimum requirements for conditional-use areas, which includes particular considerations as to their proper location to adjacent, established or intended uses, or to the planned growth of the community.

Chapter 3, along with Chapters 4 through 7, establish the criteria to classify properties into compatible use districts.

Chapter 4 Agricultural Zones. Chapter 4 identifies three divisions of agricultural zones including any area to be designated as open space, agricultural uses and land used for public parks and similar uses. After the specific zoning areas are established, this chapter provides minimum bulk zoning regulations to establish lot area, structure-to-open space density, lot dimensions, and setback and building height requirements.

For example, within an agricultural zone 2 there is a limit of one dwelling unit per 10 acres with a minimum lot area of 10 acres. Lot dimensions are required to be 400 feet wide by 400 feet deep minimum for this parcel of ground.

Chapter 5 Residential Zones. The objective of Chapter 5 is to define residential uses for a jurisdiction to utilize in arranging compatible land uses in order to achieve the maximum social and economic benefit for the community. This chapter identifies three divisions of residential zones including single-family, two-family and multiunit residential uses. Once the particular zones are established, provisions for the minimum bulk zoning regulations, such as lot area, structure-to-open-space density, lot dimensions, setback and building height requirements, are indicated. For example, based on Table 502.1, a Division 2b residential lot would restrict the overall building height to 35 feet and establish a minimum front yard of 15 feet, side yard of 5 feet and rear yard of 20 feet while requiring a minimum overall lot size of 6,000 square feet.

Chapter 6 Commercial and Commercial/Residential Zones. Chapter 6 identifies four divisions of commercial zones, including C1, which includes minor automotive repair and automotive fuel dispensing facilities; C2, which includes light commercial and group care facilities; C3, which includes amusement centers including bowling alleys, golf driving ranges, miniature golf courses, ice skating rinks, pool and billiard halls; and C4, which includes major automotive repair, manufacturing and commercial centers. This chapter also contains two divisions of commercial/residential zones that accommodate residential uses in light and medium commercial zones (Divisions 1 and 2). Once the particular zones are established, Chapter 6 provides specific minimum bulk zoning restrictions to include lot area, structure-to-open-space density, lot dimensions, and setback and building height requirements.

Chapter 7 Factory/Industrial Zones. The objective of Chapter 7 is to define factory/industrial uses for a jurisdiction to utilize in arranging compatible land uses for the social and economic welfare of the community. This chapter identifies three divisions of factory/industrial zones, including a range of factory/industrial zones from light manufacturing or industrial, such as warehouses and auto body shops (Division 1), to heavy manufacturing or industrial, such as automotive dismantling and petroleum refineries (Division 3). Once the particular zones are established, Chapter 7 provides minimum bulk zoning regulations that establish lot area, structure-to-open-space density, lot dimensions, and setback and building height requirements.

Chapter 8 General Provisions. Chapter 8 contains general zoning provisions along with requirements for elements that are common to most uses recognized by this code, to include parking stall dimensions, driveway width requirements, allowable projections into required yard spaces, landscaping and loading space size requirements. This chapter also establishes the minimum number of required off-street parking spaces for specific uses, fence height requirements specific to front, side and rear yard locations, accessory buildings and minimum separation distance requirements from accessory buildings to the main building on the same lot, maximum allowable projection encroachment into the required front and rear yards, and landscaping requirements for new buildings and additions and maintenance requirements for existing landscaping. Chapter 8 also provides for the jurisdiction to specifically review and approve the availability of essential services infrastructure for all new projects with a focus on sewer, potable water, street lighting and fire hydrants.

Chapter 9 Special Regulations. Chapter 9 recognizes two unique uses, home office and adult-use businesses, and establishes requirements to address each based on their characteristics and potential impact related to other uses/zoning districts. With respect to home occupations, Chapter 9 contains restrictions that include maximum allowable floor area for both the home occupation and the storage for same, exterior display and patron and parking allowances.

With respect to adult uses, Chapter 9 requires adult uses to obtain a conditional-use permit and contains a list of four specific location requirements for adult uses.

Chapter 10 Sign Regulations. The primary purpose of Chapter 10 is to establish the regulation for the use of signs and sign structures. This chapter addresses the various sign types, provides numerous figures that show examples of general signs, roof signs, wall signs and fascia signs, and addresses the computation methodology of sign area for code compliance. Chapter 10 also contains the general provisions that apply to sign placement, maintenance, repair and removal, as well as requirements for wall, free-standing, directional and temporary signs.

Chapter 11 Nonconforming Structures and Uses. Chapter 11 contains provisions for nonconforming structures and uses regulated under this code. The primary purpose of this chapter is to ensure that existing structures and current land use practices legally established prior to the adoption of the *International Zoning Code* are allowed to be continued. This chapter also describes the criteria that a nonconforming structure or use must meet in order to be allowed to be maintained unchanged. Specific criteria is provided for the discontinuance of a nonconforming use to include vacancy and damage. Chapter 11 also describes the restrictions on enlargements and modifications to a nonconforming structure.

Chapter 12 Conditional Uses. Chapter 12 establishes the requirements for conditional uses based on the occasional need for a use not normally permitted in a particular zoning district and due to the unique characteristics and service that use provides to the public. This chapter contains requirements for conditional-use permits, minimum documentation required to support a conditional-use property and fees. Further, Chapter 12 establishes the criteria for expiration and revocation of conditional-use permit and includes a provision that allows the applicant to submit an amendment to a conditional-use permit.

Chapter 13 Planned Unit Development. Chapter 13 identifies the code requirements for planned unit developments and describes the important role of the planning commission. The primary purpose of this chapter is to permit and encourage diversification, variation and imagination in the relationship of uses, structures, open spaces and heights of structures. It is further intended to encourage more rational and economic development with relationship to public services, and to encourage and facilitate the preservation of open lands. This chapter contains the specific conditions for planned unit developments, including area, uses, ownership and open space requirements.

Chapter 14 Referenced Standards. The code contains numerous references to standards that are used to regulate materials and methods of construction. Chapter 14 contains a comprehensive list of all standards that are referenced in the code. The standards are part of the code to the extent of the reference to the standard. Compliance with the referenced standard is necessary for compliance with this code. By providing specifically adopted standards, the construction and installation requirements necessary for compliance with the code can be readily determined. The basis for code compliance is, therefore, established and available on an equal basis to the code official, contractor, designer and owner.

Chapter 14 is organized in a manner that makes it easy to locate specific standards. It lists all of the referenced standards, alphabetically, by acronym of the promulgating agency of the standard. Each agency's standards are then listed in either alphabetical or numeric order based on the standard identification. The list also contains the title of the standard; the edition (date) of the standard referenced; any addenda included as part of the ICC adoption; and the section or sections of this code that reference the standard.

TABLE OF CONTENTS

CHAPTER 1

SCOPE AND ADMINISTRATION

User note:

About this chapter: *Chapter 1 establishes the limits of applicability of the code and describes how the code is to be applied and enforced. Chapter 1 is in two parts: Part 1—Scope and Application (Section 101) and Part 2—Administration and Enforcement (Sections 102–111). Section 101 identifies which buildings and structures come under its purview and references other I-codes as applicable.*

This code is intended to be adopted as a legally enforceable document and it cannot be effective without adequate provisions for its administration and enforcement. The provisions of Chapter 1 establish the authority and duties of the code official appointed by the authority having jurisdiction and also establish the rights and privileges of the design professional, contractor and property owner.

PART 1—SCOPE AND APPLICATION

SECTION 101 GENERAL

[A] 101.1 Title. These regulations shall be known as the *Zoning Code* of **[NAME OF JURISDICTION]**, hereinafter referred to as "this code."

[A] 101.2 Intent. The purpose of this code is to safeguard the health, property and public welfare by controlling the design, location, use or occupancy of all buildings and structures through the regulated and orderly development of land and land uses within this jurisdiction.

[A] 101.3 Scope. The provisions of this code shall apply to the construction, addition, alteration, moving, repair and use of any building, structure, parcel of land or *sign* within a jurisdiction, except work located primarily in a *public way*, public utility towers and poles and public utilities unless specifically mentioned in this code.

Where there is conflict between a general requirement and a specific requirement, the specific requirement shall be applicable. Where, in any specific case, different sections of this code specify different requirements, the more restrictive shall govern.

In fulfilling these purposes, this ordinance is intended to benefit the public as a whole and not any specific person or class of persons. Although, through the implementation, administration and enforcement of this code, benefits and detriments will be enjoyed or suffered by specific individuals, such is merely a byproduct of the overall benefit to the whole community. Therefore, unintentional breaches of the obligations of administration and enforcement imposed on the jurisdiction hereby shall not be enforceable in tort.

If any portion of this code is held invalid for any reason, the remaining herein shall not be affected.

PART 2—ADMINISTRATION AND ENFORCEMENT

SECTION 102 EXISTING BUILDINGS AND USES

[A] 102.1 General. Lawfully established buildings and uses in existence at the time of the adoption of this code shall be permitted to have their existing use or occupancy continued, provided that such continued use is not dangerous to life.

[A] 102.2 Additions, alterations or repairs. Additions, alterations or repairs shall be permitted to be made to any building or use without requiring the existing building or use to comply with the requirements of this code, provided that the addition, alteration or repair conforms to that required for a new building or use.

[A] 102.3 Maintenance. Buildings or uses, both existing and new, and all parts thereof, shall be maintained. The owner or owner's authorized agent shall be responsible for the maintenance of buildings and parcels of land. To determine compliance with this section, the code official shall be permitted to cause any structure or use to be inspected.

[A] 102.4 Moved and temporary buildings, structures and uses. Buildings or structures moved into or within the jurisdiction shall comply with the provisions of this code for new buildings and structures.

Temporary buildings, structures and uses such as reviewing stands and other miscellaneous structures, sheds, canopies or fences used for the protection of the public shall be permitted to be erected, provided that a special approval is received from the code official for a limited period of time. *Temporary buildings* or structures shall be completely removed upon the expiration of the time limit stated in the permit.

[A] 102.5 Illegal uses. Uses that were illegally established prior to the adoption of this code shall remain illegal.

SECTION 103 PLANNING COMMISSION

[A] 103.1 General. This section addresses the duties and responsibilities of a planning commission, hereafter referred to as "the commission," and other officials and agencies, with respect to the administration of this code.

[A] 103.2 Establishment of the commission. The establishment of the commission shall be in accordance with the policies and procedures as set forth in state law. The commission shall consist of the number of members as specified in state law. Additionally, one member of the legislative body shall be permitted to be appointed as liaison to the commission. Such member shall have the right to attend all meetings and take part in all discussions, but shall not vote on commission decisions.

[A] 103.3 Terms for members. The terms of office for the members of the commission shall be as set forth in state law. Members shall be permitted to be removed for cause upon written charges and after a public hearing before the legislative body of the jurisdiction, if such a hearing is requested.

[A] 103.4 Selection of members. Members shall be appointed and approved by the legislative body of the jurisdiction served. The terms of office for the commission members shall be staggered at intervals so as to provide continuity in policy and personnel. Members of the commission shall be residents of the jurisdiction served. Compensation of members shall be set by the legislative body of the jurisdiction. Any vacancy for the unexpired term of any member whose term is not completed shall be filled. A member shall continue to serve until a successor has been appointed and approved by the legislative body of the jurisdiction.

[A] 103.5 Chairperson election and rules adoption. The commission shall elect from its membership a chairperson. It shall establish and adopt rules for its organization and transaction of business and shall keep a public record of its proceedings.

[A] 103.6 Commission secretary. A secretary to assist the commission shall be appointed by the code official. The secretary shall keep minutes of the commission meetings for public record and conduct all correspondence, including the notification of decisions. The secretary shall certify records. The secretary shall prepare and submit the minutes of commission meetings to the chairperson and the commission.

[A] 103.7 Duties and powers. The duties and powers of the planning commission shall be in accordance with Sections 103.7.1 through 103.7.5.

[A] 103.7.1 Comprehensive plan. It shall be the duty of the commission, after holding public hearings, to create and recommend to the legislative body a comprehensive plan for the physical development of the jurisdiction, which shall be permitted to include areas outside its boundaries that bear consideration to the planning of the jurisdiction. The comprehensive plan shall include at least the following elements:

1. Official maps.
2. Growth and land use.
3. Commercial/industrial uses.
4. Transportation and utilities.
5. Community facilities.
6. Housing.
7. Environmental.
8. Geologic/natural hazards.

The commission shall be permitted to recommend amendments to the comprehensive plan regarding the administration or maintenance of this code.

[A] 103.7.2 Zoning code. It shall be the duty of the commission to develop and recommend to the legislative body a zoning code, in accordance with the guidelines of the comprehensive plan, establishing zones within the jurisdiction. Such a code shall be made in regards to the character of each district and the most appropriate use of land within the jurisdiction.

The commission shall make periodic reports and recommendations to the legislative body.

[A] 103.7.3 Division of land regulations. It shall be the duty of the commission to develop and certify regulations governing the division of land. Divisions of land shall be in accordance with the adopted regulations.

[A] 103.7.4 Conditional-use permits. It shall be the duty of the commission to review conditional-use permit applications.

The application shall be accompanied by maps, drawings or other documentation in support of the request. The granting of a conditional-use permit shall not exempt the applicant from compliance with other relevant provisions of related ordinances.

[A] 103.7.5 Official zoning map. The legislative body shall adopt an official zoning map for all areas included within the jurisdiction.

[A] 103.8 Appeals and hearings. Any person with standing aggrieved by any decision of the commission shall have the right to make such appeals as shall be permitted to be provided by this code or state law. Such appeals shall be based on the record.

SECTION 104 DUTIES AND POWERS OF THE ZONING CODE OFFICIAL

[A] 104.1 General. This section establishes the duties and responsibilities for the zoning code official and other officials and agencies, with respect to the administration of this code. The zoning code official and/or designee shall be referred to hereafter as "the code official."

[A] 104.2 Deputies. The code official may appoint such number of technical officers and other employees as shall be authorized from time to time. The code official shall be permitted to deputize such employees as may be necessary to carry out the functions of this code.

[A] 104.3 Reviews and approvals. The code official shall be authorized to undertake reviews, make recommendations and grant approvals as set forth in this code.

[A] 104.4 Comprehensive plan. The code official shall assist the planning commission in the development and implementation of the comprehensive plan.

[A] 104.5 Administrative reviews and permits. Administrative reviews and permits shall be in accordance with Sections 104.5.1 through 104.5.4.

[A] 104.5.1 Review of building permits. Applications for building permits and amendments thereto shall be submitted to the code official for review and approved prior to permit issuance. Each application shall include a set of building plans and all data necessary to show that the requirements of this code are met.

[A] 104.5.2 Site plan reviews. The code official shall receive all applications for site plan review and review for completeness and prepare submittals for review by the appropriate body.

[A] 104.5.3 Conditional-use permits and variances. The code official shall receive all applications for conditional uses and variances or other plans as shall be permitted or approved as required by this code, review for completeness and prepare submittals for review by the appropriate body.

[A] 104.5.4 Amendments. Requests for amendments or changes to the comprehensive plan or this code or map shall be submitted to the code official for processing.

[A] 104.6 Interpretations. The interpretation and application of the provisions of this code shall be by the code official. An appeal of an interpretation by the code official shall be submitted to the board of adjustment, who, unless otherwise provided, is authorized to interpret the code, and such interpretation shall be considered to be final.

Uses are permitted within the various zones as described in this code and as otherwise provided herein.

It is recognized that all possible uses and variations of uses that might arise cannot reasonably be listed or categorized. Mixed uses/sites or any use not specifically mentioned or about which there is any question shall be administratively classified by comparison with other uses identified in the zones described in this code. If the proposed use resembles identified uses in terms of intensity and character, and is consistent with the purpose of this code and the individual zone's classification, it shall be considered as a permitted/nonpermitted use within a general zone classification, subject to the regulations for the use it most nearly resembles. If a use does not resemble other identified allowable uses within a zone, it may be permitted as determined by the hearing body in public hearing as an amendment to this code pursuant to Section 109.3.

[A] 104.7 Liability. The code official, or designee, charged with the enforcement of this code, acting in good faith and without malice in the discharge of the duties described in this code, shall not be personally civilly or criminally liable for any damage that may accrue to persons or property as a result of an act or by reason of an act or omission in the discharge of such duties.

[A] 104.7.1 Legal defense. A suit or criminal complaint brought against the code official or employee because such act or omission performed by the code official or employee in the enforcement of any provision of such codes or other pertinent laws or ordinances implemented through the enforcement of this code or enforced by the enforcement agency shall be defended by the jurisdiction until final termination of such proceedings. Any judgment resulting therefrom shall be assumed by the jurisdiction.

This code shall not be construed to relieve from or lessen the responsibility of any person owning, operating or controlling any building or parcel of land for any damages to persons or property caused by defects, nor shall the enforcement agency or its jurisdiction be held as assuming any such liability by reason of the reviews or permits issued under this code.

[A] 104.8 Cooperation of other officials and officers. The code official shall be authorized to request, and shall receive so far as is required in the discharge of the duties described in this code, the assistance and cooperation of other officials of the jurisdiction.

SECTION 105 COMPLIANCE WITH THE CODE

[A] 105.1 General. Upon adoption of this code by the legislative body, no use, building or structure, whether publicly or privately owned, shall be constructed or authorized until the location and extent thereof conform to said plan.

SECTION 106 BOARD OF ADJUSTMENT

[A] 106.1 General. This section addresses the duties and responsibilities of a board of adjustment, hereafter referred to as "the board," and other officials and agencies, with respect to the administration of this code.

[A] 106.2 Establishment of the board. The establishment of the board shall be in accordance with the procedures and policies set forth in state law. The board shall consist of the number of members as specified in state law. Additionally, one member of the commission shall be appointed as liaison to the board. Such member shall have the right to attend all meetings and take part in all discussions, but shall not vote on board decisions.

[A] 106.3 Terms for members. The terms of office for the members of the board shall be as set forth in state law. Members shall be permitted to be removed for cause upon written charges and after a public hearing before the legislative body of the jurisdiction, if such hearing is requested.

[A] 106.4 Selection of members. Members shall be appointed and approved by the legislative body of the jurisdiction served. The terms of office shall be staggered at intervals, so as to provide continuity in policy and personnel. Members of the board shall be residents of the jurisdiction served. Compensation shall be set by the legislative body of the jurisdiction. Any vacancy for the unexpired term of any member whose term is not completed shall be filled. A member shall continue to serve until a successor has been appointed and approved by the legislative body of the jurisdiction.

[A] 106.5 Chairperson election and rules adoption. The board shall elect from its membership a chairperson. It shall establish and adopt rules for its organization and the transaction of business and shall keep a public record of its proceedings.

[A] 106.6 Board secretary. A secretary to assist the board shall be appointed by the code official. The secretary shall keep minutes of the board meetings for public record and conduct all correspondence, including the notification of decisions. The secretary shall certify records. The secretary shall prepare and submit the minutes of board meetings to the chairperson and the board.

[A] 106.7 Duties and powers. The duties and powers of the board of adjustment shall be in accordance with Sections 106.7.1 through 106.7.3.

[A] 106.7.1 Errors. The board shall have the power to hear and decide on appeals where it is alleged that there is an error in any order, requirement, decision, determination or interpretation by the code official.

[A] 106.7.2 Variances. The board shall have the power to hear and decide on appeals wherein a variance to the terms of this code is proposed. Limitations as to the board's authorization shall be as set forth in this code.

[A] 106.7.3 Variance review criteria. The board of adjustment shall be permitted to approve, approve with conditions or deny a request for a variance. Each request for a variance shall be consistent with the following criteria:

1. Limitations on the use of the property due to physical, topographical and geologic features.
2. The grant of the variance will not grant any special privilege to the property owner or the owner's authorized agent.
3. The applicant can demonstrate that without a variance there can be no reasonable use of the property.
4. The grant of the variance is not based solely on economic reasons.
5. The necessity for the variance was not created by the property owner or the owner's authorized agent.
6. The variance requested is the minimum variance necessary to allow reasonable use of the property.
7. The grant of the variance will not be injurious to the public health, safety or welfare.
8. The property subject to the variance request possesses one or more unique characteristics generally not applicable to similarly situated properties.

[A] 106.8 Use variance. The board of adjustment shall not grant a variance to allow the establishment of a use in a zoning district where such use is prohibited by the provisions of this code.

[A] 106.9 Decisions. The board shall be permitted to decide in any manner it sees fit; however, it shall not have the authority to alter or change this code or zoning map or allow as a use that would be inconsistent with the requirements of this code, provided, however, that in interpreting and applying the provisions of this code, the requirements shall be deemed to be the spirit and intent of the code and do not constitute the granting of a special privilege.

SECTION 107
HEARING EXAMINER

[A] 107.1 General. This section addresses the duties and responsibilities of a hearing examiner, hereafter referred to as the "examiner," and other officials and agencies with respect to the administration of this code.

[A] 107.2 Appointment of an examiner. The examiner shall be appointed and approved by the legislative body of the jurisdiction served. Compensation shall be set by same.

[A] 107.3 Duties and powers. The examiner shall hear and consider all applications for discretionary land rezones and use decisions as authorized by the legislative body by resolution. Such considerations shall be set for public hearing. The examiner shall be bound by the same standards of conduct as the commission and board, with respect to the administration of this code.

[A] 107.4 Decisions. The examiner shall, within 10 working days, render a decision. Notice in writing of the decision and the minutes of record shall be given to the code official for distribution as required. Decisions shall be kept in accordance with state regulations and such decisions shall be open to the public.

SECTION 108
HEARINGS, APPEALS AND AMENDMENTS

[A] 108.1 Hearings. Upon receipt of an application in proper form, the code official shall arrange to advertise the time and place of public hearing. Such advertisement shall be given by not fewer than one publication in a newspaper of general circulation within the jurisdiction. Such notice shall state the nature of the request, the location of the property, and the time and place of hearing. Reasonable effort shall be made to give notice by regular mail of the time and place of hearing to each surrounding property owner or the owner's authorized agent; the extent of the area to be notified shall be set by the code official. A notice of such hearing shall be posted in a conspicuous manner on the subject property.

[A] 108.2 Appeals. Appeals shall be in accordance with Sections 108.2.1 through 108.2.3.

[A] 108.2.1 Filing. Any person with standing, aggrieved or affected by any decision of the code official, shall be permitted to appeal to the examiner, board or commission by written request with the code official. Upon furnishing the proper information, the code official shall transmit to the examiner, board or commission all papers and pertinent data related to the appeal.

[A] 108.2.2 Time limit. An appeal shall only be considered if filed within [NUMBER OF WORKING DAYS] days after the cause arises or the appeal shall not be considered. If such an appeal is not made, the decision of the code official shall be considered to be final.

[A] 108.2.3 Stays of proceedings. An appeal stays all proceedings from further action unless there is immediate danger to public health and safety.

[A] 108.3 Amendments. This code shall be permitted to be amended, but all proposed amendments shall be submitted to the code official for review and recommendation to the commission.

[A] 108.4 Voting and notice of decision. There shall be a vote of a majority of the board and commission present in order to decide any matter under consideration. Each decision shall be entered in the minutes by the secretary. Appeals shall be kept in accordance with state regulations and such appeals shall be open to the public.

Notice in writing of the decision and the disposition of each appeal shall be given to the code official and each appellant by mail or otherwise.

SECTION 109
VIOLATIONS

[A] 109.1 Unlawful acts. It shall be unlawful for any person to erect, construct, enlarge, alter, repair, move, improve, remove, convert or demolish, equip, use, occupy, or maintain any building or land or cause or permit the same to be done in violation of this code. Where any building or parcel of land regulated by this code is being used contrary to this code, the code official shall be permitted to order such use discontinued and the structure, parcel of land, or portion thereof, vacated by notice served on any person causing such use to be continued. Such person shall discontinue the use within the time prescribed by the code official after receipt of such notice to make the structure, parcel of land, or portion thereof, comply with the requirements of this code.

SECTION 110 PERMITS AND APPROVALS

[A] 110.1 General. Departments, officials and employees which are charged with the duty or authority to issue permits or approvals shall issue no permit or approval for uses or purposes where the same would be in conflict with this code. Any permit or approval, if issued in conflict with this code, shall be null and void.

[A] 110.2 Expiration or cancellation. Each license, permit or approval issued shall expire after 180 days if no work is undertaken or such use or activity is not established, unless a different time of issuance of the license or permit is allowed in this code, or unless an extension is granted by the issuing agency prior to expiration.

Failure to comply fully with the terms of any permit, license or approval shall be permitted to be grounds for cancellation or revocation. Action to cancel any license, permit or approval shall be permitted to be taken on proper grounds by the code official. Cancellation of a permit or approval by the commission or board shall be permitted to be appealed in the same manner as its original action.

[A] 110.3 Validity of licenses, permits and approvals. For the issuance of any license, permit or approval for which the commission or board is responsible, the code official shall require that the development or use in question proceed only in accordance with the terms of such license, permit or approval, including any requirements or conditions established as a condition of issuance. Except as specifically provided for in this code and conditions of approval, the securing of one required review or approval shall not exempt the recipient from the necessity of securing any other required review or approval.

SECTION 111 FEES

[A] 111.1 Fees. A fee for services shall be charged. Fees shall be set by the jurisdiction and schedules shall be available at the office of the code official.

CHAPTER 2

DEFINITIONS

User note:

About this chapter: *Codes, by their very nature, are technical documents. Every word, term and punctuation mark can add to or change the meaning of a technical requirement. It is necessary to maintain a consensus on the specific meaning of each term contained in the code. Chapter 2 performs this function by stating clearly what specific terms mean for the purpose of the code.*

SECTION 201
GENERAL

201.1 Scope. Unless otherwise expressly stated, the following words and terms shall, for the purposes of this code, have the meanings shown in this chapter.

201.2 Interchangeability. Words used in the present tense include the future; words stated in the masculine gender include the feminine and neuter; the singular number includes the plural and the plural, the singular.

201.3 Terms defined in other codes. Where terms are not defined in this code and are defined in the *International Building Code* or the *International Mechanical Code,* such terms shall have the meanings ascribed to them as in those codes.

201.4 Terms not defined. Where terms are not defined through the methods authorized by this section, such terms shall have ordinarily accepted meanings such as the context implies.

SECTION 202
GENERAL DEFINITIONS

ABANDONED SIGN. See Section 1002.1.

ACCESSORY BUILDING. An incidental subordinate building customarily incidental to and located on the same lot occupied by the main use or building, such as a detached garage.

ACCESSORY LIVING QUARTERS. An accessory building used solely as the temporary dwelling of guests of the occupants of the premises; such dwelling having no *kitchen* facilities and not rented or otherwise used as a separate *sleeping unit.*

ACCESSORY USE. A use conducted on the same lot as the primary use of the structure to which it is related; a use that is clearly incidental to, and customarily found in connection with, such primary use.

AGRICULTURE. The tilling of the soil, raising of crops, *farm animals, livestock,* horticulture, gardening, beekeeping and aquaculture.

ALLEY. Any *public way* or thoroughfare more than 10 feet (3048 mm), but less than 16 feet (4877 mm), in width, which has been dedicated to the public for public use.

[A] ALTERATION. Any change, addition or modification in construction, occupancy or use.

AMUSEMENT CENTER. An establishment offering five or more amusement devices, including, but not limited to, coin-operated electronic games, shooting galleries, table games and similar recreational diversions within an enclosed building.

ANIMATED SIGN. See Section 1002.1.

Electrically activated. See Section 1002.1.

Environmentally activated. See Section 1002.1.

Mechanically activated. See Section 1002.1.

APARTMENT HOUSE. A residential building designed or used for three or more dwelling units.

ARCHITECTURAL PROJECTION. See Section 1002.1.

AUTOMOTIVE REPAIR, MAJOR. An establishment primarily engaged in the repair or maintenance of motor vehicles, trailers and similar large mechanical equipment, including paint, body and fender, and major engine and engine part overhaul, which is conducted within a completely enclosed building.

AUTOMOTIVE REPAIR, MINOR. An establishment primarily engaged in the repair or maintenance of motor vehicles, trailers and similar mechanical equipment, including brake, muffler, upholstery work, tire repair and change, lubrication, tune ups, and transmission work, which is conducted within a completely enclosed building.

AUTOMOTIVE SELF-SERVICE MOTOR FUEL DISPENSING FACILITY. That portion of property where flammable or combustible liquids or gases used as fuel are stored and dispensed from fixed equipment into the fuel tanks of motor vehicles by persons other than a service station attendant. Such an establishment shall be permitted to offer for sale at retail other convenience items as a clearly secondary activity and shall be permitted to include a free-standing automatic car wash.

AUTOMOTIVE SERVICE MOTOR FUEL DISPENSING FACILITY. That portion of property where flammable or combustible liquids or gases used as fuel are stored and dispensed from fixed equipment into the fuel tanks of motor vehicles. Accessory activities shall be permitted to include automotive repair and maintenance, car wash service, and food sales.

AWNING. See Section 1002.1.

AWNING SIGN. See Section 1002.1.

BACKLIT AWNING. See Section 1002.1.

BANNER. See Section 1002.1.

BANNER SIGN. See Section 1002.1.

BASEMENT. Any floor level below the first *story* in a building, except that a floor level in a building having only one floor level shall be classified as a *basement* unless such floor level qualifies as a first *story* as defined herein.

BILLBOARD. See Section 1002.1.

BOARD. The board of adjustment of the adopting jurisdiction.

[BG] BOARDING HOUSE. A dwelling containing a single *dwelling unit* and not more than 10 sleeping units, where lodging is provided with or without meals, for compensation for more than one week.

[A] BUILDING. Any structure used or intended for supporting or sheltering any use or occupancy.

BUILDING, MAIN. A building in which the principal use of the site is conducted.

BUILDING, TEMPORARY. A building used temporarily for the storage of construction materials and equipment incidental and necessary to on-site permitted construction of utilities, or other community facilities, or used temporarily in conjunction with the sale of property within a subdivision under construction.

BUILDING CODE. The *International Building Code* promulgated by the International Code Council, as adopted by the jurisdiction.

BUILDING ELEVATION. See Section 1002.1.

BUILDING HEIGHT. The vertical distance above the average existing *grade* measured to the highest point of the building. The height of a stepped or terraced building shall be the maximum height of any segment of the building.

BUILDING LINE. The perimeter of that portion of a building or structure nearest a property line, but excluding open steps, terraces, cornices and other ornamental features projecting from the walls of the building or structure.

BUSINESS OR FINANCIAL SERVICES. An establishment intended for the conduct or service or administration by a commercial enterprise, or offices for the conduct of professional or business service.

CANOPY. A roofed structure constructed of fabric or other material supported by the building or by support extending to the ground directly under the *canopy* placed so as to extend outward from the building providing a protective shield for doors, windows and other openings.

CANOPY (Attached). See Section 1002.1.

CANOPY (Free-standing). See Section 1002.1.

CANOPY SIGN. See Section 1002.1.

CHANGEABLE SIGN. See Section 1002.1.

Electrically activated. See Section 1002.1.

Manually activated. See Section 1002.1.

COMBINATION SIGN. See Section 1002.1.

COMMERCIAL, HEAVY. An establishment or business that generally uses open sales yards, outside equipment storage or outside activities that generate noise or other impacts considered incompatible with less-intense uses. Typical businesses in this definition are lumber yards, construction specialty services, heavy equipment suppliers or building contractors.

COMMERCIAL, LIGHT. An establishment or business that generally has retail or wholesale sales, office uses, or services, which do not generate noise or other impacts considered incompatible with less-intense uses. Typical businesses in this definition are retail stores, offices, catering services or restaurants.

COMMERCIAL CENTER, COMMUNITY. A completely planned and designed commercial development providing for the sale of general merchandise and/or convenience goods and services. A *community commercial center* shall provide for the sale of general merchandise, and may include a variety store, discount store or supermarket.

COMMERCIAL CENTER, CONVENIENCE. A completely planned and designed commercial development providing for the sale of general merchandise and/or convenience goods and services. A *convenience commercial center* shall provide a small cluster of convenience shops or services.

COMMERCIAL CENTER, NEIGHBORHOOD. A completely planned and designed commercial development providing for the sale of general merchandise and/or convenience goods and services. A *neighborhood commercial center* shall provide for the sales of convenience goods and services, with a supermarket as the principal tenant.

COMMERCIAL CENTER, REGIONAL. A completely planned and designed commercial development providing for the sale of general merchandise and/or convenience goods and services. A regional center shall provide for the sale of general merchandise, apparel, furniture, home furnishings, and other retail sales and services, in full depth and variety.

COMMERCIAL RETAIL SALES AND SERVICES. Establishments that engage in the sale of general retail goods and accessory services. Businesses within this definition include those that conduct sales and storage entirely within an enclosed structure (with the exception of occasional outdoor "sidewalk" promotions); businesses specializing in the sale of either general merchandise or convenience goods.

COMPREHENSIVE PLAN. The declaration of purposes, policies and programs for the development of the jurisdiction.

CONDITIONAL USE. A use that would become harmonious or compatible with neighboring uses through the application and maintenance of qualifying conditions.

CONDOMINIUM. A single-dwelling unit in a multiunit dwelling or structure, that is separately owned and may be combined with an undivided interest in the common areas and facilities of the property.

CONGREGATE RESIDENCE. Any building or portion thereof that contains facilities for living, sleeping and sanitation as required by this code, and may include facilities for eating and cooking for occupancy by other than a family. A *congregate residence* shall be permitted to be a shelter, convent, monastery, dormitory, fraternity or sorority house, but does not include jails, hospitals, nursing homes, hotels or lodging houses.

COPY. See Section 1002.1.

COURT. A space, open and unobstructed to the sky, located at or above *grade* level on a lot and bounded on three or more sides by walls of a building.

DAY CARE, FAMILY. The keeping for part-time care and/or instruction, whether or not for compensation, of six or less children at any one time within a dwelling, not including members of the family residing on the premises.

DAY CARE, GROUP. An establishment for the care and/or instruction, whether or not for compensation, of seven or more persons at any one time. Child nurseries, preschools and adult care facilities are included in this definition.

DENSITY. The number of dwelling units that are allowed on an area of land, which area of land shall be permitted to include dedicated streets contained within the development.

DEVELOPMENT COMPLEX SIGN. See Section 1002.1.

DIRECTIONAL SIGN. See Section 1002.1.

DOUBLE-FACED SIGN. See Section 1002.1.

DRIVEWAY. A private access road, the use of that is limited to persons residing, employed, or otherwise using or visiting the parcel in which it is located.

DWELLING, MULTIPLE UNIT. A building or portion thereof designed for occupancy by three or more families living independently in which they may or may not share common entrances and/or other spaces. Individual dwelling units may be owned as condominiums, or offered for rent.

DWELLING, SINGLE FAMILY. A detached *dwelling unit* with *kitchen* and sleeping facilities, designed for occupancy by one family.

DWELLING, TWO FAMILY. A building designed or arranged to be occupied by two families living independently, with the structure having only two dwelling units.

[BG] DWELLING UNIT. A single unit providing complete, independent living facilities for one or more persons, including permanent provisions for living, sleeping, eating, cooking and sanitation.

EASEMENT. That portion of land or property reserved for present or future use by a person or agency other than the legal fee owner(s) of the property. The *easement* shall be permitted to be for use under, on or above said lot or lots.

ELECTRIC SIGN. See Section 1002.1.

ELECTRONIC MESSAGE SIGN OR CENTER. See Section 1002.1.

EXTERIOR SIGN. See Section 1002.1.

FACE OF BUILDING, PRIMARY. The wall of a building fronting on a street or right-of-way, excluding any appurtenances such as projecting fins, columns, pilasters, canopies, marquees, showcases or decorations.

FARM ANIMALS. Animals other than household pets that shall be permitted to, where permitted, be kept and maintained for commercial production and sale and/or family food production, education or recreation. *Farm animals* are identified by these categories: large animals, for example, horses and cattle; medium animals, for example, sheep and goats; or small animals, for example, rabbits, chinchillas, chickens, turkeys, pheasants, geese, ducks and pigeons.

FASCIA SIGN. See "Wall or fascia *sign*," Section 1002.1.

FLASHING SIGN. See "Animated *sign*, electrically activated," Section 1002.1.

FLOOR AREA, GROSS. The sum of the horizontal areas of floors of a building measured from the exterior face of exterior walls or, if appropriate, from the center line of dividing walls; this includes courts and decks or porches where covered by a roof.

FLOOR AREA, NET. The *gross floor area* exclusive of vents, shafts, courts, elevators, stairways, exterior walls and similar facilities.

FREE-STANDING SIGN. See Section 1002.1.

FRONTAGE. The width of a lot or parcel abutting a public right-of-way measured at the front property line.

FRONTAGE (Building). See Section 1002.1.

FRONTAGE (Property). See Section 1002.1.

GARAGE, PRIVATE. A building or a portion of a building not more than 1,000 square feet (92.9 m^2) in area, in which only private or pleasure-type motor vehicles used by the tenants of the building or buildings on the premises are stored or kept.

GRADE (Adjacent Ground Elevation). The lowest point of elevation of the existing surface of the ground, within the area between the building and a line 5 feet (1524 mm) from the building.

GROUND SIGN. See "Free-standing *sign*," Section 1002.1.

GROUP CARE FACILITY. A facility, required to be licensed by the state, which provides training, care, supervision, treatment and/or rehabilitation to the aged, disabled, those convicted of crimes, or those suffering the effects of drugs or alcohol; this does not include day care centers, *family day care* homes, foster homes, schools, hospitals, jails or prisons.

[BG] HABITABLE SPACE (Room). Space in a structure for living, sleeping, eating or cooking. Bathrooms, toilet compartments, closets, halls, storage or utility space, and similar areas are not considered *habitable space*.

HOME OCCUPATION. The partial use of a home for commercial or nonresidential uses by a resident thereof, which is subordinate and incidental to the use of the dwelling for residential purposes.

HOSPITAL. An institution designed for the diagnosis, treatment and care of human illness or infirmity and providing health services, primarily for inpatients, and including as related facilities, laboratories, outpatient departments, training facilities and staff offices.

HOUSEHOLD PETS. Dogs, cats, rabbits, birds, and the like, for family use only (noncommercial) with cages, pens, etc.

ILLUMINATED SIGN. See Section 1002.1.

INDUSTRIAL OR RESEARCH PARK. A tract of land developed in accordance with a master site plan for the use of a family of industries and their related commercial uses, and that is of sufficient size and physical improvement to protect surrounding areas and the general community and to ensure a harmonious integration into the neighborhood.

INTERIOR SIGN. See Section 1002.1.

[A] JURISDICTION. The governmental unit that has adopted this code.

KITCHEN. Any room or portion of a room within a building designed and intended to be used for the cooking or preparation of food.

LANDSCAPING. The finishing and adornment of unpaved *yard* areas. Materials and treatment generally include naturally growing elements such as grass, trees, shrubs and flowers. This treatment shall be permitted to include the use of logs, rocks, fountains, water features and contouring of the earth.

LEGISLATIVE BODY. The political entity of the adopting jurisdiction.

LIVESTOCK. Includes, but is not limited to, horses, bovine animals, sheep, goats, swine, reindeer, donkeys, mules and any other hoofed animals.

[A] LOT. A portion or parcel of land considered as a unit.

MANSARD. See Section 1002.1.

MANUFACTURING, HEAVY. Other types of manufacturing not included in the definitions of *light manufacturing* and *medium manufacturing*.

MANUFACTURING, LIGHT. The manufacturing, compounding, processing, assembling, packaging or testing of goods or equipment, including research activities, conducted entirely within an enclosed structure, with no outside storage, serviced by a modest volume of trucks or vans and imposing a negligible impact on the surrounding environment by noise, vibration, smoke, dust or pollutants.

MANUFACTURING, MEDIUM. The manufacturing, compounding, processing, assembling, packaging or testing of goods or equipment within an enclosed structure or an open *yard* that is capable of being screened from neighboring properties, serviced by a modest volume of trucks or other vehicles.

MARQUEE. See "*Canopy* (Attached)," Section 1002.1.

MARQUEE SIGN. See "*Canopy sign*," Section 1002.1.

MENU BOARD. See Section 1002.1.

MORTUARY, FUNERAL HOME. An establishment in which the dead are prepared for burial or cremation. The facility shall be permitted to include a chapel for the conduct of funeral services and spaces for funeral services and informal gatherings, and/or display of funeral equipment.

MOTEL, HOTEL. Any building containing six or more sleeping units intended or designed to be used, or that are used, rented or hired out to be occupied, or that are occupied for sleeping purposes by guests.

MULTIPLE-FACED SIGN. See Section 1002.1.

NONCONFORMING LOT. A lot where the width, area or other dimension did not conform to the regulations when this code became effective.

NONCONFORMING SIGN. A sign or sign structure or portion thereof lawfully existing at the time this code became effective, which does not now conform.

NONCONFORMING STRUCTURE. A building or structure or portion thereof lawfully existing at the time this code became effective, which was designed, erected or structurally altered for a use that does not conform to the zoning regulations of the zone in which it is located.

NONCONFORMING USE. See "Use, nonconforming."

OFF-PREMISE SIGN. See "Outdoor advertising *sign*," Section 1002.1.

ON-PREMISE SIGN. See Section 1002.1.

OPEN SPACE. Land areas that are not occupied by buildings, structures, parking areas, streets, alleys or required yards. *Open space* shall be permitted to be devoted to *landscaping*, preservation of natural features, patios, and recreational areas and facilities.

OUTDOOR ADVERTISING SIGN. See Section 1002.1.

PARAPET. See Section 1002.1.

PARK. A public or private area of land, with or without buildings, intended for outdoor active or passive recreational uses.

PARKING LOT. An open area, other than a street, used for the parking of automobiles.

PARKING SPACE, AUTOMOBILE. A space within a building or private or public parking lot, exclusive of driveways, ramps, columns, office and work areas, for the parking of an automobile.

[A] PERSON. An individual, heirs, executors, administrators or assigns, and includes a firm, partnership or corporation, its or their successors or assigns, or the agent of any of the aforesaid.

PLANNED UNIT DEVELOPMENT (PUD). A residential or commercial development guided by a total design plan in which one or more of the zoning or subdivision regulations, other than use regulations, shall be permitted to be waived or varied to allow flexibility and creativity in site and building design and location, in accordance with general guidelines.

PLOT PLAN. A plot of a lot, drawn to scale, showing the actual measurements, the size and location of any existing buildings or buildings to be erected, the location of the lot in relation to abutting streets, and other such information.

POLE SIGN. See "Free-standing *sign*," Section 1002.1.

POLITICAL SIGN. See Section 1002.1.

POOLS (SWIMMING), HOT TUBS AND SPAS.

Above-ground/on-ground pool. See "Private swimming pool."

Barrier. A fence, a wall, a building wall, the wall of an above-ground swimming pool or a combination thereof, which completely surrounds the swimming pool and obstructs access to the swimming pool.

Hot tub. See "Private swimming pool."

In-ground pool. See "Private swimming pool."

Power safety cover. A pool cover that is placed over the water area, and is opened and closed with a motorized mechanism activated by a control switch.

Private swimming pool. Any structure that contains water over 24 inches (610 mm) in depth and that is used, or intended to be used, for swimming or recreational bathing in connection with an occupancy in Use Group R-3 and that is available only to the family and guests of the householder. This includes in-ground, above-ground, and on-ground swimming pools, hot tubs and spas.

Private swimming pool, indoor. Any private *swimming pool* that is totally contained within a private structure and surrounded on all four sides by walls of said structure.

Private swimming pool, outdoor. Any private swimming pool that is not an indoor pool.

Public swimming pool. Any swimming pool other than a private swimming pool.

Spa. See "Private swimming pool."

PORTABLE SIGN. See Section 1002.1.

PROJECTING SIGN. See Section 1002.1.

PUBLIC IMPROVEMENT. Any drainage ditch, storm sewer or drainage facility, sanitary sewer, water main, roadway, parkway, sidewalk, pedestrian way, tree, lawn, off-street parking area, lot improvement, or other facility for which the local government may ultimately assume the responsibility for maintenance and operation, or for which the local government responsibility is established.

PUBLIC SERVICES. Uses operated by a unit of government to serve public needs, such as police (with or without jail), fire service, ambulance, judicial *court* or government offices, but not including public utility stations or maintenance facilities.

PUBLIC UTILITY STATION. A structure or facility used by a public or quasi-public utility agency to store, distribute, generate electricity, gas, telecommunications, and related equipment, or to pump or chemically treat water. This does not include storage or treatment of sewage, solid waste or hazardous waste.

[A] PUBLIC WAY. Any street, *alley* or other parcel of land open to the outside air, that has been deeded, dedicated or otherwise permanently appropriated to the public for public use and has a clear width and height of not less than 10 feet (3048 mm).

QUASI-PUBLIC. Essentially a public use, although under private ownership or control.

QUORUM. A majority of the authorized members of a board or commission.

REAL ESTATE SIGN. See Section 1002.1.

RECREATION, INDOOR. An establishment providing completely enclosed recreation activities. Accessory uses shall be permitted to include the preparation and serving of food and/or the sale of equipment related to the enclosed uses. Included in this definition shall be bowling, roller skating or ice skating, billiards, pool, motion picture theatres, and related amusements.

RECREATION, OUTDOOR. An area free of buildings except for restrooms, dressing rooms, equipment storage, maintenance buildings, open-air pavilions and similar structures used primarily for recreational activities.

RECYCLING FACILITY. Any location whose primary use is where waste or scrap materials are stored, bought, sold, accumulated, exchanged, packaged, disassembled or handled, including, but not limited to, scrap metals, paper, rags, tires and bottles, and other such materials.

[A] REGISTERED DESIGN PROFESSIONAL. An individual who is registered or licensed to practice their respective design profession as defined by the statutory requirements of the professional registration laws of the state or jurisdiction in which the project is to be constructed.

REHABILITATION CENTER (Halfway House). An establishment where the primary purpose is the rehabilitation of persons. Such services include drug and alcohol rehabilitation, assistance to emotionally and mentally disturbed persons, and halfway houses for prison parolees and juveniles.

RELIGIOUS, CULTURAL AND FRATERNAL ACTIVITY. A use or building owned or maintained by organized religious organizations or nonprofit associations for social, civic or philanthropic purposes, or the purpose for which persons regularly assemble for worship.

RENOVATION. Interior or exterior remodeling of a structure, other than ordinary repair.

RESTAURANT. An establishment that sells prepared food for consumption. Restaurants shall be classified as follows:

Restaurant, fast food. An establishment that sells food already prepared for consumption, packaged in paper, styrofoam or similar materials, and may include drive-in or drive-up facilities for ordering.

Restaurant, general. An establishment that sells food for consumption on or off the premises.

Restaurant, take-out. An establishment that sells food only for consumption off the premises.

REVOLVING SIGN. See Section 1002.1.

ROOF LINE. See Section 1002.1.

ROOF SIGN. See Section 1002.1.

SCHOOL, COMMERCIAL. A school establishment to provide for the teaching of *industrial*, clerical, managerial or artistic skills. This definition applies to schools that are owned and operated privately for profit and that do not offer a complete educational curriculum (for example, beauty school or modeling school).

SETBACK. The minimum required distance between the property line and the *building line*.

SIGN. An advertising message, announcement, declaration, demonstration, display, illustration, insignia, surface or space erected or maintained in view of the observer thereof for identification, advertisement or promotion of the interests of any person, entity, product or service, including the *sign* structure, supports, lighting system and any attachments, ornaments or other features used to draw the attention of observers.

SIGN (Chapter 10). See Section 1002.1.

SIGN AREA. See Section 1002.1.

SIGN COPY. See Section 1002.1.

SIGN FACE. See Section 1002.1.

SIGN STRUCTURE. See Section 1002.1.

SIGNS, COMMUNITY. Temporary, on- or off-premises signs, generally made of a woven material or durable synthetic materials primarily attached to or hung from light poles or on buildings. These signs are solely of a decorative, festive and/or informative nature announcing activities, promotions or events with seasonal or traditional themes having broad community interest, and which are sponsored or supported by a jurisdiction-based nonprofit organization.

SITE PLAN. A plan that outlines the use and development of any tract of land.

[BG] SLEEPING UNIT. A room or space in which people sleep, which can also include permanent provisions for living, eating and either sanitation or *kitchen* facilities, but not both. Such rooms and spaces that are also part of a *dwelling unit* are not sleeping units.

[BG] STORY. That portion of a building included between the upper surface of any floor and the upper surface of the floor next above, except that the topmost *story* shall be that portion of a building included between the upper surface of the topmost floor and the ceiling or roof above. If the finished floor level directly above a usable or unused under-floor space is more than 6 feet (1829 mm) above *grade* as defined herein for more than 50 percent of the total perimeter or is more than 12 feet (3658 mm) above *grade* as defined herein at any point, such usable or unused under-floor space shall be considered to be a *story*.

STREET. Any thoroughfare or *public way* not less than 16 feet (4877 mm) in width that has been dedicated.

STREET, PRIVATE. A right-of-way or *easement* in private ownership, not dedicated or maintained as a public street, that affords the principal means of access to two or more sites.

[A] STRUCTURE. That which is built or constructed.

SUBDIVISION. The division of a tract, lot or parcel of land into two or more lots, plats, sites or other divisions of land.

TEMPORARY SIGN. See Section 1002.1.

THEATER. A building used primarily for the presentation of live stage productions, performances or motion pictures.

UNDER CANOPY SIGN OR UNDER MARQUEE SIGN. See Section 1002.1.

USE. The activity occurring on a lot or parcel for which land or a building is arranged, designed or intended, or for which land or a building is or may be occupied, including all accessory uses.

USE, CHANGE OF. The change within the classified use of a structure or premise.

USE, NONCONFORMING. A use that lawfully occupied a building or land at the time this code became effective, which has been lawfully continued and which does not now conform to the use regulations.

USE, PRINCIPAL. A use that fulfills a primary function of a household, establishment, institution or other entity.

USE, TEMPORARY. A use that is authorized by this code to be conducted for a fixed period of time. Temporary uses are characterized by such activities as the sale of agricultural products, contractors' offices and equipment sheds, fireworks, carnivals, flea markets, and garage sales.

V SIGN. See Section 1002.1.

VARIANCE. A deviation from the height, bulk, setback, parking or other dimensional requirements established by this code.

WALL OR FASCIA SIGN. See Section 1002.1.

WAREHOUSE, WHOLESALE OR STORAGE. A building or premises in which goods, merchandise or equipment are stored for eventual distribution.

WINDOW SIGN. See Section 1002.1.

YARD. An open, unoccupied space on a lot, other than a *court*, that is unobstructed from the ground upward by buildings or structures, except as otherwise provided in this code.

YARD, FRONT. A *yard* extending across the full width of the lot, the depth of which is the minimum horizontal distance between the front lot line and a line parallel thereto.

YARD, REAR. A *yard* extending across the full width of the lot, the depth of which is the minimum horizontal distance between the rear lot line or ordinary high water line and a line parallel thereto.

YARD, SIDE. An open, unoccupied space on the same lot with the building and between the *building line* and the side lot line, or to the ordinary high water line.

CHAPTER 3
USE DISTRICTS

User note:

About this chapter: *Chapter 3 identifies classifications for typical zoning districts and provides for the application of minimum district areas in order to regulate and restrict the uses and locations of buildings and to regulate the minimum required areas for yards, courts and open spaces.*

This chapter also establishes requirements for jurisdictional zoning maps and minimum requirements for conditional-use areas, which include particular considerations as to their location to established or intended uses, or to the planned growth of the community.

SECTION 301
DISTRICT CLASSIFICATIONS

301.1 Classification. In order to classify, regulate and restrict the locations of uses and locations of buildings designated for specific areas; and to regulate and determine the areas of yards, courts and other open spaces within or surrounding such buildings, property is hereby classified into districts as prescribed in this chapter.

SECTION 302
MINIMUM AREAS FOR ZONING DISTRICTS

302.1 Minimum areas. The minimum areas that may constitute a separate or detached part of any of the following zoning districts on the zoning map or subsequent amendments to said zoning map shall be as shown in Table 302.1. Where a nonresidential district is directly across the street from or abuts the district with the same or less restrictive classification, the area of the land directly across the street or abutting the property may be included in the calculations in meeting the minimum district size requirements.

TABLE 302.1
MINIMUM AREAS FOR ZONING DISTRICTS

ZONING DISTRICT	MINIMUM AREA[a] OF THE DISTRICT
A, Division 1	No minimum
A, Division 2	[JURISDICTION TO INSERT NUMBER]
A, Division 3	[JURISDICTION TO INSERT NUMBER]
C, Division 1	[JURISDICTION TO INSERT NUMBER]
C, Division 2	[JURISDICTION TO INSERT NUMBER]
C, Division 3	[JURISDICTION TO INSERT NUMBER]
C, Division 4	[JURISDICTION TO INSERT NUMBER]
CR, Division 1	[JURISDICTION TO INSERT NUMBER]
CR, Division 2	[JURISDICTION TO INSERT NUMBER]
FI, Division 1	[JURISDICTION TO INSERT NUMBER]
FI, Division 2	[JURISDICTION TO INSERT NUMBER]
FI, Division 3	[JURISDICTION TO INSERT NUMBER]
R, Division 1	[JURISDICTION TO INSERT NUMBER]
R, Division 2	[JURISDICTION TO INSERT NUMBER]
R, Division 3	[JURISDICTION TO INSERT NUMBER]

For SI: 1 acre = 4047 m^2.

a. The adopting jurisdiction should fill in with appropriate land areas expressed in acres.

SECTION 303
ZONING MAP

303.1 General. The boundaries of each zoning district are to be indicated on the official zoning map as approved by the legislative authority. Said map and subsequent amendments thereto shall be considered to be a part of this code.

SECTION 304
ANNEXED TERRITORY

304.1 Classification. Any territory hereafter annexed shall automatically, upon such annexation, be classified as R, Division 1a, residential district, and be subject to all conditions and regulations applicable to property in such district.

SECTION 305
CONDITIONAL USES

305.1 General. The principal objective of this zoning code is to provide for an orderly arrangement of compatible buildings and land uses, and for the property location of all types of uses required for the social and economic welfare of the community. To accomplish this objective, each type and kind of use is classified as permitted in one or more of the various use districts established by this code. However, in addition to those uses specifically classified and permitted in each district, there are certain additional uses that it may be necessary to allow because of the unusual characteristics of the service they provide the public. These conditional uses require particular considerations as to their proper location to adjacent, established or intended uses, or to the planned growth of the community. The conditions controlling the locations and operation of such special uses are established by the applicable sections of this code.

CHAPTER 4
AGRICULTURAL ZONES

User note:

About this chapter: *Chapter 4 identifies three divisions of agricultural zones including open space, agricultural uses and land used for public parks and similar uses. After the specific zoning areas are established, this chapter provides minimum bulk zoning regulations to establish lot area, structure-to-open-space density, lot dimensions, and setback and building height requirements.*

SECTION 401
AGRICULTURAL ZONES DEFINED

401.1 Agricultural zone. Allowable agricultural (A) zone uses shall be:

Division 1. Any designated *open space* as set forth in this code.

Division 2. Any agricultural use, including, but not limited to, dwellings, maintenance/storage buildings and other such uses necessary for the principal use.

Division 3. Any public park land or other similar recreational use, including, but not limited to, amusement rides, office buildings, retail buildings and dwellings necessary for the maintenance of the principal use.

SECTION 402
BULK REGULATIONS

402.1 General. The minimum area, setbacks, *density* and maximum height shall be as prescribed in Table 402.1.

TABLE 402.1
AGRICULTURAL (A) ZONE BULK REGULATIONS (in feet, unless noted otherwise)[a]

ZONE DIVISION	MINIMUM LOT AREA (acres)	MAXIMUM DENSITY (units/acre)	LOT DIMENSIONS		SETBACK REQUIREMENTS			MAXIMUM BUILDING HEIGHT[b]
			Minimum lot width	Minimum lot depth	Minimum front yard	Minimum side yard	Minimum rear yard	
1	20	1 dwelling unit/20 acres	600	600	30	15	60	35
2	10	1 dwelling unit/10 acres	400	400	30	15	60	35
3	5	1 dwelling unit/5 acres	250	250	30	15	60	35

For SI: 1 foot = 304.8 mm, 1 acre = 4047 m^2.

a. Open spaces and parks can be of a reduced size, if approved.

b. Access storage structures, windmills and similar structures shall be permitted to exceed the maximum height where approved by the code official.

CHAPTER 5
RESIDENTIAL ZONES

User note:

About this chapter: *Chapter 5 identifies three divisions of residential zones including single-family, two-family and multiunit residential uses. Once the particular zones are established, provisions for the minimum bulk zoning regulations, such as lot area, structure-to-open-space density, lot dimensions, setback and building height requirements, are indicated. The objective of this chapter is to define residential uses for a jurisdiction to utilize in arranging compatible land uses in order to achieve the maximum social and economic benefit for the community.*

SECTION 501
RESIDENTIAL ZONES DEFINED

501.1 Residential zone. Allowable residential (R) zone uses shall be:

Division 1. The following uses are permitted in an R, Division 1 zone:

Single-family dwellings, publicly owned and operated parks, recreation centers, swimming pools and playgrounds, police and fire department stations, public and governmental services, public libraries, schools and colleges (excluding colleges or trade schools operated for profit), public parking lots, private garages, buildings accessory to the above permitted uses (including private garages and accessory living quarters), and temporary buildings.

Division 2. The following uses are permitted in an R, Division 2 zone:

Any use permitted in R, Division 1 zones and two-family dwellings.

Division 3. The following uses are permitted in an R, Division 3 zone:

All uses permitted in R, Division 2 zones, multiple-unit dwellings, such as apartment houses, boarding houses, condominiums and congregate residences.

SECTION 502
BULK REGULATIONS

502.1 General. The minimum area, setbacks, *density* and maximum height shall be as prescribed in Table 502.1.

TABLE 502.1
RESIDENTIAL (R) ZONE BULK REGULATIONS (in feet, unless noted otherwise)

DIVISION		MINIMUM LOT AREA/ SITE (square feet)	MAXIMUM DENSITY (dwelling unit/acre)	LOT DIMENSIONS		SETBACK REQUIREMENTS			MAXIMUM BUILDING HEIGHT[a]
				Minimum lot width	Minimum lot depth	Minimum front yard	Minimum side yard	Minimum rear yard	
1	a	35,000	1	125	150	25	10	30	35
	b	20,000	2	100	125	20	10	25	35
	c	10,000	4	75	100	20	5	25	35
	d	6,000	6	60	90	15	5	20	35
2	a	10,000	4	60	70	20	5	20	35
	b	6,000	6	60	70	15	5	20	35
3	a	6,000	8	60	70	15	5	20	35
	b	6,000	12	60	70	15	5	20	35

For SI: 1 foot = 304.8 mm, 1 square foot = 0.0929 m^2, 1 acre = 4047 m^2.

a. Accessory towers, satellite dishes and similar structures shall be permitted to exceed the maximum height where approved by the code official.

CHAPTER 6

COMMERCIAL AND COMMERCIAL/RESIDENTIAL ZONES

User note:

About this chapter: *Chapter 6 identifies four divisions of commercial zones, which includes minor automotive repair and automotive fuel dispensing facilities; light commercial and group care facilities; amusement centers including bowling alleys, golf driving ranges, miniature golf courses, ice skating rinks, pool and billiard halls; major automotive repair; manufacturing and commercial centers. This chapter also contains two divisions of commercial/residential zones that accommodate residential uses in light and medium commercial zones. Once the particular zones are established, this chapter provides specific minimum bulk zoning restrictions to include lot area, structure-to-open-space density, lot dimensions, and setback and building height requirements.*

SECTION 601 COMMERCIAL AND COMMERCIAL/ RESIDENTIAL ZONES DEFINED

601.1 Commercial and commercial/residential zones. Allowable commercial (C) zone and commercial/residential (CR) zone uses shall be:

C Zone

Division 1. The following uses are permitted in a C, Division 1 zone:

Minor automotive repair, automotive motor fuel dispensing facilities, automotive self-service motor fuel dispensing facilities, business or financial services, convenience and neighborhood commercial centers (excluding wholesale sales), family and *group day care* facilities, libraries, mortuary and funeral homes, public and governmental services, police and fire department stations, places of religious worship, public utility stations, and restaurants.

Division 2. The following uses are permitted in a C, Division 2 zone:

Any uses permitted in C, Division 1 zones, and *light commercial* (excluding wholesale sales), group care facilities, physical fitness centers, religious, cultural and fraternal activities, rehabilitation centers, and schools and colleges operated for profit (including commercial, vocational and trade schools).

Division 3. The following uses are permitted in a C, Division 3 zone:

Any uses permitted in C, Division 2 zones, and amusement centers (including bowling alleys, golf driving ranges, miniature golf courses, ice rinks, pool and billiard halls, and similar recreational uses), automotive sales, building material supply sales (wholesale and retail), cultural institutions (such as museums and art galleries), community commercial centers (including wholesale and retail sales), health and medical institutions (such as hospitals), hotels and motels (excluding other residential occupancies), commercial printing and publishing, taverns and cocktail lounges, indoor theaters, and self-storage warehouses.

Division 4. The following uses are permitted in a C, Division 4 zone:

Any uses permitted in C, Division 3 zones, and *major automotive repair*, commercial bakeries, regional commercial centers (including wholesale and retail sales), plastic products design, molding and assembly, small metal products design, casting, fabricating, and processing, manufacture and finishing, storage yards, and wood products manufacture and finishing.

CR Zone

Permitted (commercial/residential) (CR) zone uses shall be:

Division 1. The following uses are permitted in a CR, Division 1 zone:

Any use permitted in a C, Division 1 zone, and residential use permitted, except in the story or *basement* abutting street *grade*.

Division 2. The following uses are permitted in a CR, Division 2 zone:

Any use permitted in a C, Division 2 zone, and residential use permitted, except in the story or *basement* abutting street *grade*.

SECTION 602 BULK REGULATIONS

602.1 General. The minimum area, setbacks, *density* and maximum height shall be as prescribed in Table 602.1.

TABLE 602.1
COMMERCIAL (C) AND COMMERCIAL/RESIDENTIAL (CR) ZONES BULK REGULATIONS
(in feet, unless noted otherwise)

DIVISION	MINIMUM LOT AREA (square feet)	MAXIMUM DENSITY (units/acre)	LOT DIMENSIONS		SETBACK REQUIREMENTS			MAXIMUM BUILDING HEIGHT[a]
			Minimum lot width	Minimum lot depth	Minimum front yard	Minimum side yard	Minimum rear yard	
1	6,000	12	30	70	0	0	0	30
2	Not Applicable	Not Applicable	30	70	0	0	0	40
3	Not Applicable	Not Applicable	75	100	0	0	0	50
4	Not Applicable	Not Applicable	75	100	0	0	0	50

For SI: 1 foot = 304.8 mm, 1 square foot = 0.0929 m^2, 1 acre = 4047 m^2.

a. Accessory towers, satellite dishes and similar structures shall be permitted to exceed the listed heights where approved by the code official.

CHAPTER 7

FACTORY/INDUSTRIAL ZONES

User note:

About this chapter: *Chapter 7 identifies three divisions of factory/industrial zones, including a range of factory/industrial zones from light manufacturing or industrial, such as warehouses and auto body shops to heavy manufacturing or industrial, such as automotive dismantling and petroleum refineries. Once the particular zones are established, this chapter provides minimum bulk zoning regulations that establish lot area, structure-to-open-space density, lot dimensions, and setback and building height requirements. The objective of this chapter is to define factory/industrial uses for a jurisdiction to utilize in arranging compatible land uses for the social and economic welfare of the community.*

SECTION 701 FACTORY/INDUSTRIAL ZONES DEFINED

701.1 FI zones. Allowable factory/*industrial* (FI) zone uses shall be:

Division 1. Any light-manufacturing or *industrial* use, such as warehouses, research or testing laboratories, product distribution centers, woodworking shops, auto body shops, furniture assembly, dry cleaning plants, places of religious worship, public and governmental services, machine shops, and boat building storage yards.

Division 2. Any use permitted in the FI, Division 1 zone and stadiums and arenas, indoor swap meets, breweries, liquid fertilizer manufacturing, carpet manufacturing, monument works, and a regional recycling center.

Division 3. Any use permitted in the FI, Division 2 zone and auto-dismantling yards, alcohol manufacturing, cotton gins, paper manufacturing, quarries, salt works, petroleum refining, and other similar uses.

SECTION 702 BULK REGULATIONS

702.1 General. The minimum area, setbacks, *density* and maximum height shall be as prescribed in Table 702.1.

TABLE 702.1
FACTORY/INDUSTRIAL (FI) ZONE BULK REGULATIONS
(in feet, unless noted otherwise)

DIVISION	MINIMUM LOT AREA (square feet)	MAXIMUM DENSITY (units/acre)	LOT DIMENSIONS		SETBACK REQUIREMENTS			MAXIMUM BUILDING HEIGHT[a]
			Minimum lot width	Minimum lot depth	Minimum front yard	Minimum side yard	Minimum rear yard	
1	Not Applicable	Not Applicable	50	75	0	0	0	60
2	Not Applicable	Not Applicable	75	100	0	0	0	80
3	Not Applicable	Not Applicable	100	150	0	0	0	80

For SI: 1 foot = 304.8 mm, 1 square foot = 0.0929 m^2, 1 acre = 4047 m^2.

a. Accessory towers, satellite dishes and similar structures shall be permitted to exceed the maximum height where approved by the code official.

CHAPTER 8
GENERAL PROVISIONS

User note:

About this chapter: *Chapter 8 contains general zoning provisions along with requirements for parking stall dimensions, driveway width requirements, allowable projections into required yard spaces, landscaping and loading space size requirements. This chapter also establishes the minimum number of off-street parking spaces, fence height requirements, accessory buildings, maximum allowable projection encroachments, and landscaping requirements for new buildings and additions. This chapter also provides for the jurisdiction to review and approve the availability of essential services, such as sewer, potable water, street lighting and fire hydrants for all new projects.*

SECTION 801
OFF-STREET PARKING

801.1 General. Off-street parking shall be provided in compliance with this chapter where any building is erected, altered, enlarged, converted or increased in size or capacity.

801.2 Parking space requirements. Parking spaces shall be in accordance with Sections 801.2.1 through 801.2.4.

801.2.1 Required number. The off-street parking spaces required for each use permitted by this code shall be not less than that found in Table 801.2.1, provided that any fractional parking space be computed as a whole space.

TABLE 801.2.1
OFF-STREET PARKING SCHEDULE

USE	NUMBER OF PARKING SPACES REQUIRED
Assembly	1 per 300 gross square feet
Dwelling unit	2 per dwelling unit
Health club	1 per 100 gross square feet
Hotel/motel	1 per sleeping unit plus 1 per 500 square feet of common area
Industry	1 per 500 square feet
Medical office	1 per 200 gross square feet
Office	1 per 300 gross square feet
Restaurant	1 per 100 gross square feet
Retail	1 per 200 gross square feet
School	1 per 3.5 seats in assembly rooms plus 1 per faculty member
Warehouse	1 per 500 gross square feet

For SI: 1 square foot = 0.0929 m^2.

801.2.2 Combination of uses. Where there is a combination of uses on a lot, the required number of parking spaces shall be the sum of that found for each use.

801.2.3 Location of lot. The parking spaces required by this code shall be provided on the same lot as the use or where the exclusive use of such is provided on another lot not more than 500 feet (152 m) radially from the subject lot within the same or less-restrictive zoning district.

801.2.4 Accessible spaces. Accessible parking spaces and passenger loading zones shall be provided in accordance with the building code. Passenger loading zones shall be designed and constructed in accordance with ICC A117.1.

801.3 Parking stall dimension. Parking stall dimensions shall be in accordance with Sections 801.3.1 and 801.3.2.

801.3.1 Width. A minimum width of 9 feet (2743 mm) shall be provided for each parking stall.

Exceptions:

1. Compact parking stalls shall be not less than 8 feet (2438 mm) wide.
2. Parallel parking stalls shall be not less than 8 feet (2438 mm) wide.
3. The width of a parking stall shall be increased 10 inches (254 mm) for obstructions located on either side of the stall within 14 feet (4267 mm) of the access aisle.
4. Accessible parking spaces shall be designed in accordance with ICC A117.1.

801.3.2 Length. A minimum length of 20 feet (6096 mm) shall be provided for each parking stall.

Exceptions:

1. Compact parking stalls shall be not less than 18 feet (5486 mm) in length.
2. Parallel parking stalls shall be not less than 22 feet (6706 mm) in length.

801.4 Design of parking facilities. The design of parking facilities shall be in accordance with Sections 801.4.1 through 801.4.7.

801.4.1 Driveway width. Every parking facility shall be provided with one or more access driveways, the width of which shall be the following:

1. Private driveways not less than 9 feet (2743 mm).
2. Commercial driveways:
   - 2.1. Twelve feet (3658 mm) for one-way enter/exit.
   - 2.2. Twenty-four feet (7315 mm) for two-way enter/exit.

801.4.2 Driveway and ramp slopes. The maximum slope of any *driveway* or ramp shall not exceed 20 percent. Transition slopes in driveways and ramps shall be provided in accordance with the standards set by the code official and the jurisdiction's engineer.

801.4.3 Stall access. Each required parking stall shall be individually and easily accessed. Automobiles shall not be required to back onto any public street or sidewalk to leave any parking stall where such stalls serve more than two dwelling units or other than residential uses. Portions of a public lot or garage shall be accessible to other portions thereof without requiring the use of any public street.

801.4.4 Compact-to-standard stall ratio. The maximum ratio of compact stalls to standard stalls in any parking area shall not exceed 1 to 2.

801.4.5 Screening. A 3-foot-high (914 mm) buffer at the public way shall be provided for all parking areas of five or more parking spaces.

801.4.6 Striping. Parking stalls shall be striped.

> **Exception:** A *private garage* or parking area for the exclusive use of a single-family dwelling.

801.4.7 Lighting. Lights illuminating a parking area shall be designed and located so as to reflect away from any street and adjacent property.

SECTION 802
FENCE HEIGHTS

802.1 General. Fence and retaining wall heights in required yards shall not exceed those found in Table 802.1.

TABLE 802.1
MAXIMUM FENCE HEIGHTS

YARDS	HEIGHT (feet)
Front	3.5
Rear	6.0
Side	
Lot side	6.0
Street side	3.5

For SI: 1 foot = 304.8 mm.

SECTION 803
LOCATION OF ACCESSORY BUILDINGS

803.1 General. Accessory buildings shall occupy the same lot as the main use or building.

803.2 Separation from main building. Accessory buildings shall be separated from the main building by 10 feet (3048 mm).

803.3 Private garages. An accessory building used as a *private garage* shall be permitted to be located in the rear yard or side yard provided that setbacks are maintained and the structures do not encroach into any recorded easements. The building shall be permitted to be located in the front yard of a sloping lot if the lot has more than a 10-foot (3048 mm) difference in elevation from midpoint of the front lot line to a point 50 feet (15 240 mm) away midway between the side lot lines.

803.4 Storage buildings. Accessory buildings used for storage or other similar use shall be permitted to be located in any portion of the rear yard or side yard. Storage buildings shall not be located in the front yard.

SECTION 804
ALLOWABLE PROJECTIONS INTO YARDS

804.1 General. Eaves, cornices or other similar architectural features shall be permitted to project into a required yard not more than 12 inches (305 mm). Chimneys shall be permitted to project not more than 2 feet (610 mm), provided that the width of any side yard is not reduced to less than 30 inches (762 mm).

804.2 Front yards. Open, unenclosed ramps, porches, platforms or landings, not covered by a roof, shall be permitted to extend not more than 6 feet (1829 mm) into the required front yard, provided that such porch does not extend above the first level and is not more than 6 feet (1829 mm) above *grade* at any point.

804.3 Rear yards. Windows shall be permitted to project into a required rear yard not more than 6 inches (152 mm).

SECTION 805
LANDSCAPING REQUIREMENTS

805.1 General. *Landscaping* is required for all new buildings and additions over 500 square feet (46.5 m^2) as defined in this section. Said *landscaping* shall be completed within 1 year from the date of occupancy of the building.

805.2 Front yards. Front yards required by this code shall be completely landscaped, except for those areas occupied by access driveways, walls and structures.

805.3 Street-side side yards. Flanking street-side side yards shall be completely landscaped, except for those areas occupied by utilities, access driveways, paved walks, walls and structures.

805.4 Maintenance. Live *landscaping* required by this code shall be properly maintained. Dead or dying *landscaping* shall be replaced immediately and all sodded areas mowed, fertilized and irrigated on a regular basis.

SECTION 806
LOADING SPACES

806.1 General. Loading spaces shall be provided on the same lot for every building in the C or FI zones. No loading space is required if prevented by an existing lawful building.

806.2 Size. Each loading space shall have a clear height of 14 feet (4267 mm) and shall be directly accessible through a usable door not less than 3 feet (914 mm) in width and 6 feet, 8 inches (2032 mm) high. The minimum area of a loading space shall be 400 square feet (37.2 m^2) and the minimum dimensions shall be 20 feet (6096 mm) long and 10 feet (3048 mm) deep.

SECTION 807
PASSAGEWAYS

807.1 Residential entrances. There shall be a passageway leading from the public way to the exterior entrance of each *dwelling unit* in every residential building of not less than 10 feet (3048 mm) in width. The passageway shall be increased by 2 feet (610 mm) for each story over two.

807.2 Separation between buildings. There shall be not less than 10 feet (3048 mm) of clear space between every main building and accessory building on a lot. There shall be not less than 20 feet (6096 mm) of clear space between every residential building and another main building on the same lot.

807.3 Location of passageways. Passageways shall be permitted to be located in that space set aside for required yards. Passageways shall be open and unobstructed to the sky and shall be permitted to have such projections as allowed for yards, provided that the users of said passageway have a clear walkway to the public way. Any space between buildings or passageways that has less width than that prescribed herein shall not be further reduced.

SECTION 808
APPROVAL FOR AND
AVAILABILITY OF ESSENTIAL SERVICES

808.1 General. Projects that require the additional use of new facilities or essential services, such as sewers, storm drains, fire hydrants, potable water, public streets, street lighting and similar services, shall obtain such approval as required by the agency providing such service prior to project approval.

Nonavailability of essential services shall be permitted to be grounds for denying permits for additional development until such services are available. The jurisdiction is not obligated to extend or supply essential services if capacity is not available. If capacity is available, the extension of services shall be by and at the cost of the developer, unless the jurisdiction agrees otherwise. Service extensions shall be designed and installed in full compliance with the jurisdiction's standards for such service, and shall be subject to review, permit and inspection as required by other policies or ordinances of the jurisdiction.

CHAPTER 9

SPECIAL REGULATIONS

User note:

About this chapter: *Chapter 9 contains requirements for home occupations that include maximum allowable floor area for the home occupation and its related storage, exterior display and parking allowances, as well as requirements for adult uses that include obtaining a conditional-use permit and specific location restrictions. Chapter 9 intends to establish requirements to address home occupations and adult uses based on their characteristics and potential impact related to other uses/zoning districts.*

SECTION 901 HOME OCCUPATIONS

901.1 General. Home occupations shall be permitted in all zones, provided that the *home occupation* is clearly and obviously subordinate to the main use or *dwelling unit* for residential purposes. Home occupations shall be conducted wholly within the primary structure on the premises.

901.2 Conditions.

1. The *home occupation* shall not exceed 15 percent of the floor area of the primary structure.
2. Other than those related by blood, marriage or adoption, not more than one person shall be employed in the *home occupation.*
3. Inventory and supplies shall not occupy more than 50 percent of the area permitted to be used as a *home occupation.*
4. There shall be no exterior display or storage of goods on said premises.
5. Home occupations involving beauty shops or barber shops shall require a conditional-use permit.
6. Sales and services to patrons shall be arranged by appointment and scheduled so that not more than one patron vehicle is on the premises at the same time.
7. Two additional parking spaces shall be provided on the premises, except only one need be provided if the *home occupation* does not have an employee. Said parking shall comply with the parking requirements in Chapter 8.

SECTION 902 ADULT USES

902.1 General. A conditional-use permit shall be obtained for all adult-use businesses.

902.2 Provisions.

1. Adult-use businesses shall not be located within 1,000 feet (305 m) of a park, school, day care center, library or religious or cultural activity.
2. Adult-use businesses shall not be located within 500 feet (152 m) of any other adult-use business or any agricultural or residential zone boundary.
3. Such distances shall be measured in a straight line without regard to intervening structures, topography and zoning.
4. Said business shall be located in FI zones and shall not be permitted as a *home occupation.*

CHAPTER 10

SIGN REGULATIONS

User note:

About this chapter: *The primary purpose of Chapter 10 is to establish the regulation for the use of signs and sign structures, including general signs, roof signs, wall signs and fascia signs. This chapter also contains the general provisions that apply to sign placement, maintenance, repair and removal, as well as requirements for wall, free-standing, directional and temporary signs.*

SECTION 1001 PURPOSE

1001.1 Purpose. The purpose of this chapter is to protect the safety and orderly development of the community through the regulation of signs and sign structures.

SECTION 1002 DEFINITIONS

1002.1 Definitions. The following words and terms shall, for the purposes of this chapter and as used elsewhere in this code, have the meanings shown herein.

ABANDONED SIGN. A sign structure that has ceased to be used, and the owner intends no longer to have used, for the display of sign copy, or as otherwise defined by state law.

ANIMATED SIGN. A sign employing actual motion or the illusion of motion. Animated signs, which are differentiated from changeable signs as defined and regulated by this code, include the following types:

Electrically activated. Animated signs producing the illusion of movement by means of electronic, electrical or electro-mechanical input and/or illumination capable of simulating movement through employment of the characteristics of one or both of the classifications noted in Items 1 and 2 as follows:

1. Flashing. Animated signs or animated portions of signs where the illumination is characterized by a repetitive cycle in which the period of illumination is either the same as or less than the period of nonillumination. For the purposes of this ordinance, flashing will not be defined as occurring if the cyclical period between on-off phases of illumination exceeds 4 seconds.
2. Patterned illusionary movement. Animated signs or animated portions of signs where the illumination is characterized by simulated movement through alternate or sequential activation of various illuminated elements for the purpose of producing repetitive light patterns designed to appear in some form of constant motion.

Environmentally activated. Animated signs or devices motivated by wind, thermal changes or other natural environmental input. Includes spinners, pinwheels, pennant strings, and/or other devices or displays that respond to naturally occurring external motivation.

Mechanically activated. Animated signs characterized by repetitive motion and/or rotation activated by a mechanical system powered by electric motors or other mechanically induced means.

ARCHITECTURAL PROJECTION. Any projection that is not intended for occupancy and that extends beyond the face of an exterior wall of a building, but that does not include signs as defined herein. See also "Awning;" "Backlit awning;" and "*Canopy*, Attached and Free-standing."

AWNING. An architectural projection or shelter projecting from and supported by the exterior wall of a building and composed of a covering of rigid or nonrigid materials and/or fabric on a supporting framework that may be either permanent or retractable, including such structures that are internally illuminated by fluorescent or other light sources.

AWNING SIGN. A sign displayed on or attached flat against the surface or surfaces of an awning. See also "Wall or fascia sign."

BACKLIT AWNING. An awning with a translucent covering material and a source of illumination contained within its framework.

BANNER. A flexible substrate on which copy or graphics may be displayed.

BANNER SIGN. A sign utilizing a banner as its display surface.

BILLBOARD. See "Off-premise sign" and "Outdoor advertising sign."

BUILDING ELEVATION. The entire side of a building, from ground level to the roofline, as viewed perpendicular to the walls on that side of the building.

CANOPY (Attached). A multisided overhead structure or architectural projection supported by attachments to a building on one or more sides and either cantilevered from such building or supported by columns at additional points. The surface(s) and/or soffit of an attached *canopy* may be illuminated by means of internal or external sources of light. See also "Marquee."

CANOPY (Free-standing). A multisided overhead structure supported by columns, but not enclosed by walls. The surface(s) and or soffit of a free-standing *canopy* may be illuminated by means of internal or external sources of light.

CANOPY SIGN. A sign affixed to the visible surface(s) of an attached or free-standing *canopy*. For reference, see Section 1003.

CHANGEABLE SIGN. A sign with the capability of content change by means of manual or remote input, including signs that are:

Electrically activated. Changeable sign where the message copy or content can be changed by means of remote electrically energized on-off switching combinations of alphabetic or pictographic components arranged on a display surface. Illumination may be integral to the components, such as characterized by lamps or other light-emitting devices; or it may be from an external light source designed to reflect off the changeable component display. See also "Electronic message sign or center."

Manually activated. Changeable sign where the message copy or content can be changed manually.

COMBINATION SIGN. A sign that is supported partly by a pole and partly by a building structure.

COPY. Those letters, numerals, figures, symbols, logos and graphic elements comprising the content or message of a sign, excluding numerals identifying a street address only.

DEVELOPMENT COMPLEX SIGN. A free-standing sign identifying a multiple-occupancy development, such as a shopping center or planned *industrial* park, that is controlled by a single owner or landlord, approved in accordance with Section 1009.2 of this chapter.

DIRECTIONAL SIGN. Any sign that is designed and erected for the purpose of providing direction and/or orientation for pedestrian or vehicular traffic.

DOUBLE-FACED SIGN. A sign with two faces, back to back.

ELECTRIC SIGN. Any sign activated or illuminated by means of electrical energy.

ELECTRONIC MESSAGE SIGN OR CENTER. An electrically activated changeable sign where the variable message capability can be electronically programmed.

EXTERIOR SIGN. Any sign placed outside a building.

FASCIA SIGN. See "Wall or fascia *sign*."

FLASHING SIGN. See "Animated *sign*, electrically activated."

FREE-STANDING SIGN. A sign principally supported by a structure affixed to the ground, and not supported by a building, including signs supported by one or more columns, poles or braces placed in or on the ground. For visual reference, see Section 1003.

FRONTAGE (Building). The length of an exterior building wall or structure of a single premise orientated to the public way or other properties that it faces.

FRONTAGE (Property). The length of the property line(s) of any single premise along either a public way or other properties on which it borders.

GROUND SIGN. See "Free-standing *sign*."

ILLUMINATED SIGN. A sign characterized by the use of artificial light, either projecting through its surface(s) (internally illuminated); or reflecting off its surface(s) (externally illuminated).

INTERIOR SIGN. Any sign placed within a building, but not including "window signs" as defined by this ordinance. Interior signs, with the exception of window signs as defined, are not regulated by this chapter.

MANSARD. An inclined decorative roof-like projection that is attached to an exterior building facade.

MARQUEE. See "*Canopy* (attached)."

MARQUEE SIGN. See "*Canopy sign*."

MENU BOARD. A free-standing sign orientated to the drive-through lane for a restaurant that advertises the menu items available from the drive-through window, and which has not more than 20 percent of the total area for such a sign utilized for business identification.

MULTIPLE-FACED SIGN. A sign containing three or more faces.

OFF-PREMISE SIGN. See "Outdoor advertising *sign*."

ON-PREMISE SIGN. A sign erected, maintained or used in the outdoor environment for the purpose of the display of messages appurtenant to the use of, products sold on, or the sale or lease of, the property on which it is displayed.

OUTDOOR ADVERTISING SIGN. A permanent sign erected, maintained or used in the outdoor environment for the purpose of the display of commercial or noncommercial messages not appurtenant to the use of, products sold on, or the sale or lease of, the property on which it is displayed.

PARAPET. The extension of a building facade above the line of the structural roof.

POLE SIGN. See "Free-standing *sign*."

POLITICAL SIGN. A temporary sign intended to advance a political statement, cause or candidate for office. A legally permitted outdoor advertising sign shall not be considered to be a political sign.

PORTABLE SIGN. Any *sign* not permanently attached to the ground or to a building or building surface.

PROJECTING SIGN. A *sign* other than a wall sign that is attached to or projects more than 18 inches (457 mm) from a building face or wall or from a structure where the primary purpose is other than the support of a sign. For visual reference, see Section 1003.

REAL ESTATE SIGN. A temporary *sign* advertising the sale, lease or rental of the property or premises on which it is located.

REVOLVING SIGN. A *sign* that revolves 360 degrees (6.28 rad) about an axis. See also "Animated sign, mechanically activated."

ROOF LINE. The top edge of a peaked roof or, in the case of an extended facade or parapet, the uppermost point of said facade or parapet.

ROOF SIGN. A *sign* mounted on, and supported by, the main roof portion of a building, or above the uppermost edge of a parapet wall of a building and that is wholly or partially supported by such a building. Signs mounted on mansard facades, pent eaves and architectural projections such as canopies or marquees shall not be considered to be roof signs. For a visual reference, and a comparison of differences between roof and fascia signs, see Section 1003.

SIGN. Any device visible from a public place that displays either commercial or noncommercial messages by means of graphic presentation of alphabetic or pictorial symbols or representations. Noncommercial flags or any flags displayed from flagpoles or staffs will not be considered to be signs.

SIGN AREA. The area of the smallest geometric figure, or the sum of the combination of regular geometric figures, that comprise the sign face. The area of any double-sided or "V" shaped *sign* shall be the area of the largest single face only. The area of a sphere shall be computed as the area of a circle. The area of all other multiple-sided signs shall be computed as 50 percent of the sum of the area of all faces of the *sign*.

SIGN COPY. Those letters, numerals, figures, symbols, logos and graphic elements comprising the content or message of a *sign*, exclusive of numerals identifying a street address only.

SIGN FACE. The surface on, against or through which the *sign* copy is displayed or illustrated, not including structural supports, architectural features of a building or sign structure, nonstructural or decorative trim, or any areas that are separated from the background surface on which the sign copy is displayed by a distinct delineation, such as a reveal or border. See Section 1003.

1. In the case of panel or cabinet-type signs, the sign face shall include the entire area of the sign panel, cabinet or face substrate on which the *sign* copy is displayed or illustrated, but not open space between separate panels or cabinets.
2. In the case of *sign* structures with routed areas of sign copy, the *sign* face shall include the entire area of the surface that is routed, except where interrupted by a reveal, border, or a contrasting surface or color.
3. In the case of signs painted on a building, or individual letters or graphic elements affixed to a building or structure, the *sign* face shall comprise the sum of the geometric figures or combination of regular geometric figures drawn closest to the edge of the letters or separate graphic elements comprising the *sign* copy, but not the open space between separate groupings of sign copy on the same building or structure.
4. In the case of *sign* copy enclosed within a painted or illuminated border, or displayed on a background contrasting in color with the color of the building or structure, the *sign* face shall comprise the area within the contrasting background, or within the painted or illuminated border.

SIGN STRUCTURE. Any structure supporting a sign.

TEMPORARY SIGN. A sign intended to display either commercial or noncommercial messages of a transitory or temporary nature. Portable signs or any sign not permanently embedded in the ground, or not permanently affixed to a building or *sign* structure that is permanently embedded in the ground, are considered temporary signs.

UNDER CANOPY SIGN OR UNDER MARQUEE SIGN. A sign attached to the underside of a *canopy* or marquee.

V SIGN. Signs containing two faces of approximately equal size, erected on common or separate structures, positioned in a "V" shape with an interior angle between faces of not more than 90 (1.57 rad) degrees with the distance between the sign faces not exceeding 5 feet (1524 mm) at their closest point.

WALL OR FASCIA SIGN. A *sign* that is in any manner affixed to any exterior wall of a building or structure and that projects not more than 18 inches (457 mm) from the building or structure wall, including signs affixed to architectural projections from a building provided that the copy area of such signs remains on a parallel plane to the face of the building facade or to the face or faces of the architectural projection to which it is affixed. For a visual reference and a comparison of differences between wall or fascia signs and roof signs, see Section 1003.

WINDOW SIGN. A *sign* affixed to the surface of a window with its message intended to be visible to and readable from the public way or from adjacent property.

SECTION 1003
GENERAL SIGN TYPES

1003.1 General. Sign types and the computation of *sign* area shall be as depicted in Figures 1003.1(1) through 1003.1(4).

SECTION 1004
GENERAL PROVISIONS

1004.1 Conformance to codes. Any sign hereafter erected shall conform to the provisions of this ordinance and the provisions of the *International Building Code* and of any other ordinance or regulation within this jurisdiction.

1004.2 Signs in rights-of-way. Signs other than an official traffic sign or similar sign shall not be erected within 2 feet (610 mm) of the lines of any street, or within any public way, unless specifically authorized by other ordinances or regulations of this jurisdiction or by specific authorization of the code official.

1004.3 Projections over public ways. Signs projecting over public walkways shall be permitted to do so only subject to the projection and clearance limits either defined herein or, if not so defined, at a minimum height of 8 feet (2438 mm) from *grade* level to the bottom of the sign. Signs, architectural projections or *sign* structures projecting over vehicular access areas must conform to the minimum height clearance limitations imposed by the jurisdiction for such structures.

1004.4 Traffic visibility. Signs or sign structures shall not be erected at the intersection of any street in such a manner as to obstruct free and clear vision, nor at any location where by its position, shape or color it may interfere with or obstruct the view of or be confused with any authorized traffic sign, signal or device.

1004.5 Computation of frontage. If a premises contains walls facing more than one property line or encompasses property frontage bounded by more than one street or other property usages, the sign area(s) for each building wall or property frontage will be computed separately for each building wall or property line facing a different frontage. The sign area(s) thus calculated shall be permitted to then be applied to permitted signs placed on each separate wall or property line frontage.

1004.6 Animation and changeable messages. Animated signs, except as prohibited in Section 1006, are permitted in commercial and *industrial* zones only. Changeable signs,

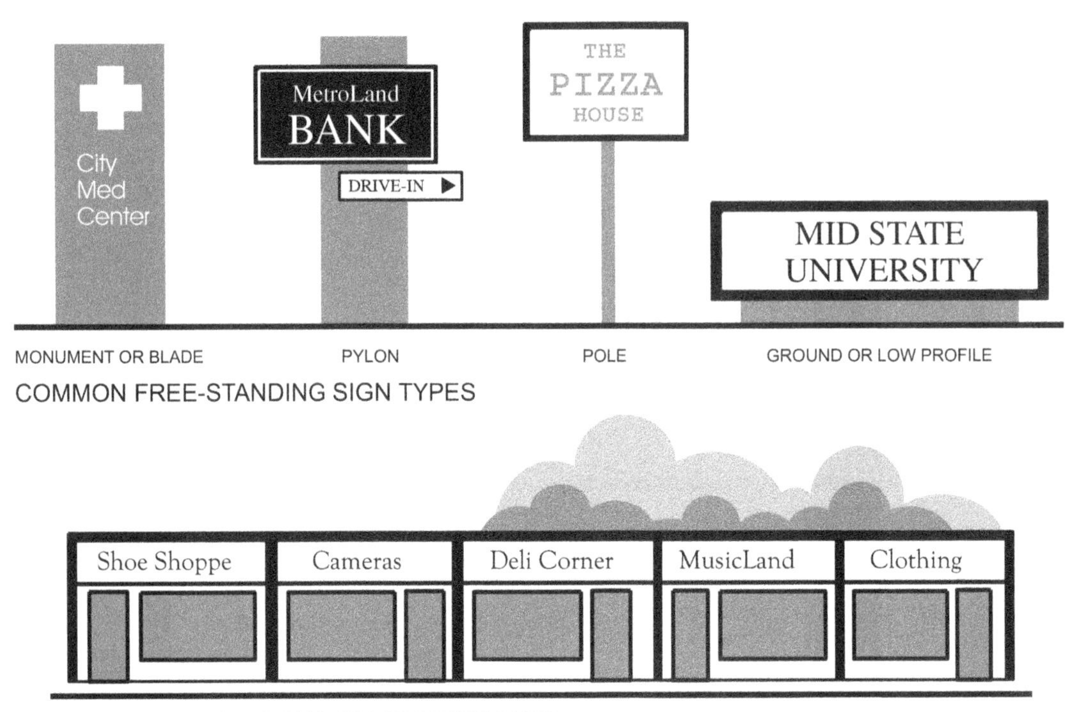

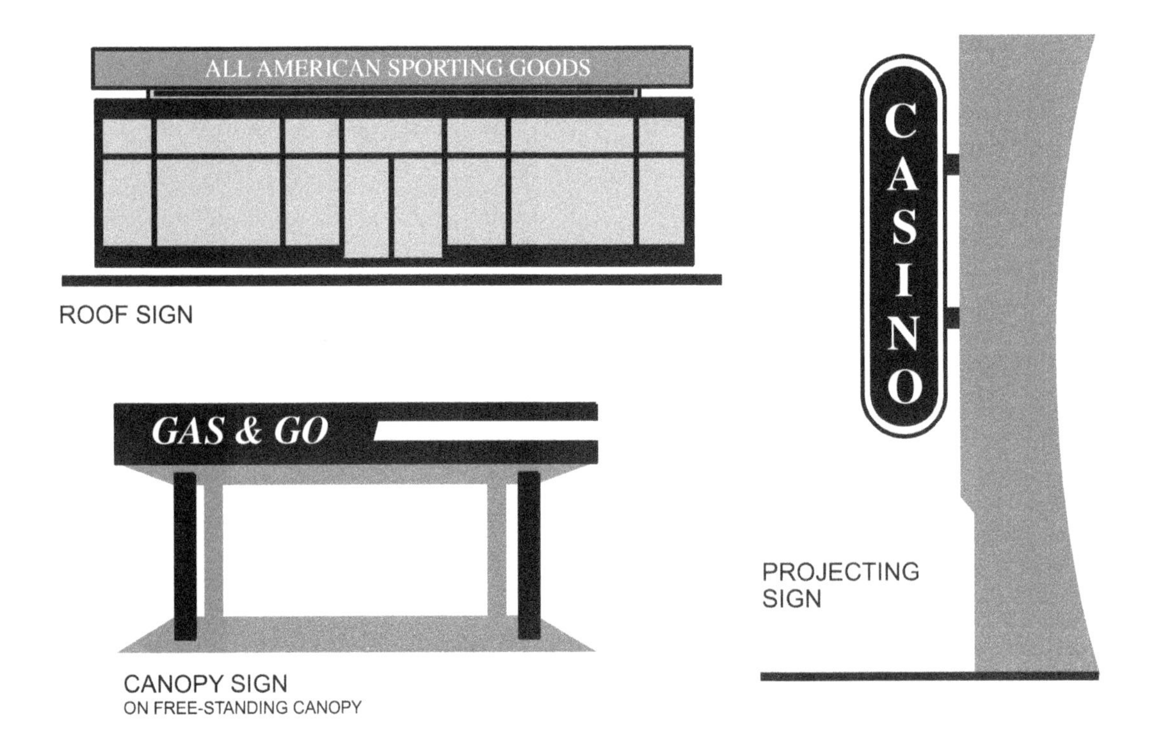

FIGURE 1003.1(1)
GENERAL SIGN TYPES

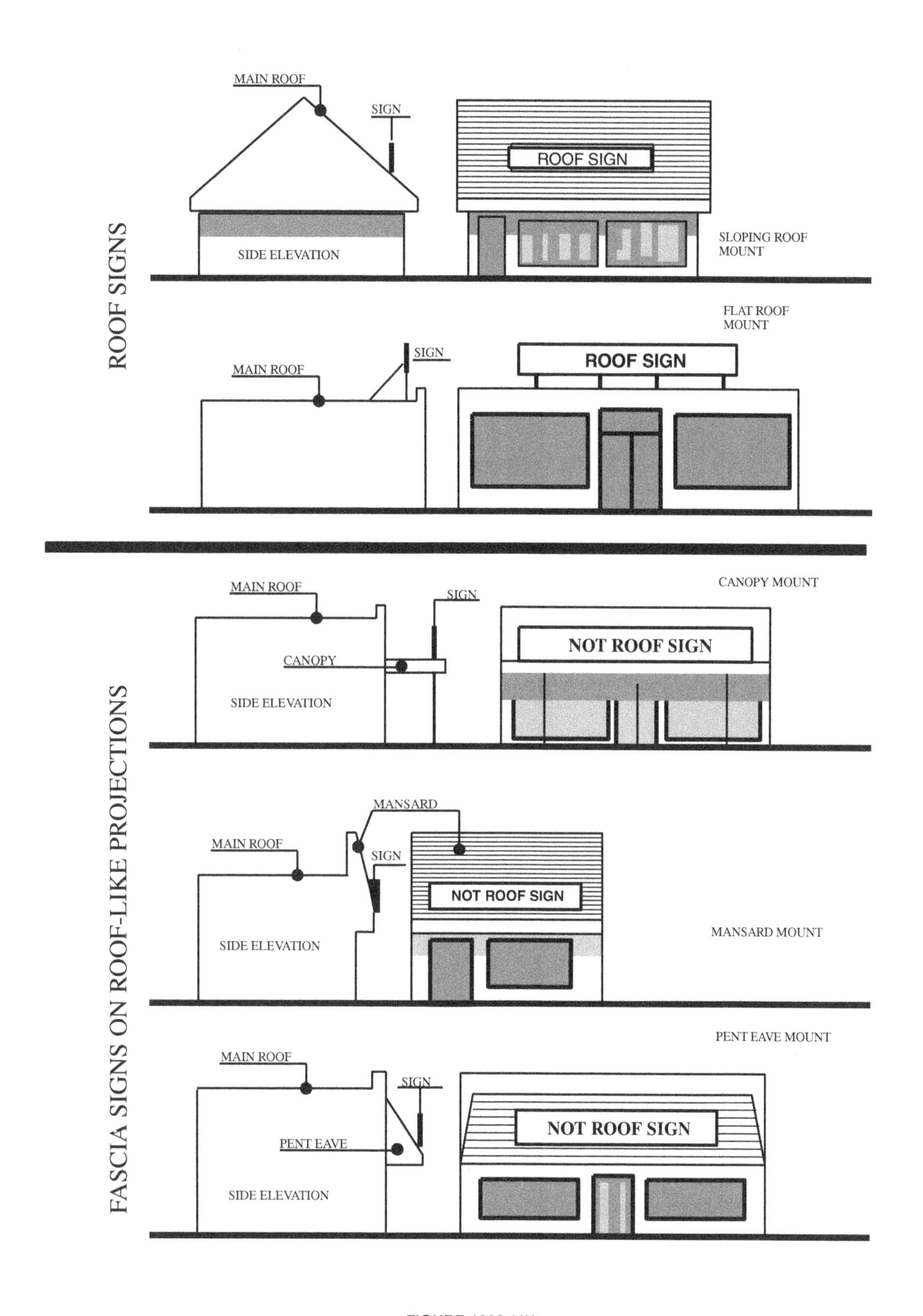

FIGURE 1003.1(2)
COMPARISON—ROOF AND WALL OR FASCIA SIGNS

SIGN STRUCTURES

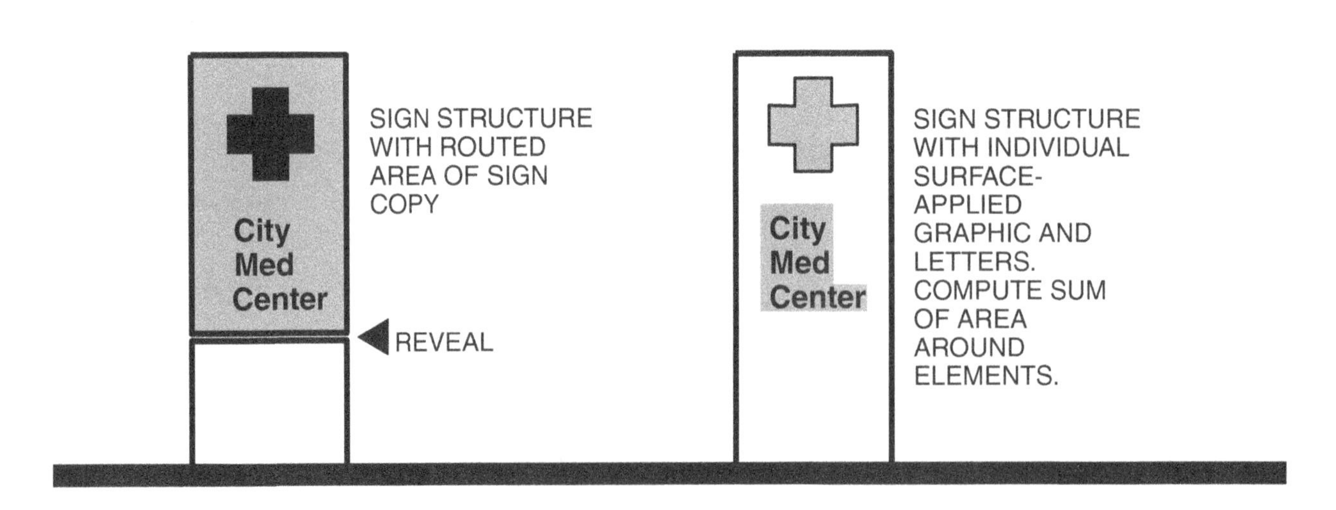

Notes: Sum of shaded areas only represents sign area. Sign constructed with panels or cabinets.

FIGURE 1003.1(3)
SIGN AREA—COMPUTATION METHODOLOGY

METROLAND BANK
Drive-In Branch

COMPUTE AREA
INSIDE DEFINED
BORDER OR
INSIDE
CONTRASTING
COLOR AREA.

METROLAND BANK
Drive-In Branch

COMPUTE SUM OF
AREAS OF INDIVIDUAL
ELEMENTS ON WALL
OR STRUCTURE.

IN COMPUTING AREA FOR UPPER-
AND LOWER-CASE LETTERING,
INCLUDE ASCENDERS OR
DESCENDERS, BUT NOT BOTH.
CALCULATE SUPER ASCENDERS
SEPARATELY AS INDICATED.

Notes: Sum of shaded areas only represents sign area for code compliance purposes. Examples of signs consisting of individual letters, elements or logos placed on building walls or structures.

FIGURE 1003.1(4)
SIGN AREA—COMPUTATION METHODOLOGY

manually activated, are permitted for nonresidential uses in all zones. Changeable signs, electrically activated, are permitted in all nonresidential zones.

1004.7 Maintenance, repair and removal. Every sign permitted by this ordinance shall be kept in good condition and repair. Where any sign becomes insecure, in danger of falling or is otherwise deemed unsafe by the code official, or if any sign shall be unlawfully installed, erected or maintained in violation of any of the provisions of this ordinance, the owner thereof or the person or firm using same shall, upon written notice by the code official forthwith in the case of immediate danger, and in any case within not more than 10 days, make such sign conform to the provisions of this ordinance, or shall remove it. If within 10 days the order is not complied with, the code official shall be permitted to remove or cause such sign to be removed at the expense of the owner and/or the user of the sign.

1004.8 Obsolete sign copy. Any *sign* copy that no longer advertises or identifies a use conducted on the property on which said *sign* is erected must have the sign copy covered or removed within 30 days after written notification from the code official; and upon failure to comply with such notice, the code official is hereby authorized to cause removal of such sign copy, and any expense incident thereto shall be paid by the owner of the building, structure or ground on which the sign is located.

1004.9 Nonconforming signs. Any sign legally existing at the time of the passage of this ordinance that does not conform in use, location, height or size with the regulations of the zone in which such *sign* is located, shall be considered to be a legal nonconforming use or structure and shall be permitted to continue in such status until such time as it is either abandoned or removed by its owner, subject to the following limitations:

1. Structural alterations, enlargement or re-erection are permissible only where such alterations will not increase the degree of nonconformity of the signs.
2. Any legal nonconforming *sign* shall be removed or rebuilt without increasing the existing height or area if it is damaged, or removed if allowed to deteriorate to the extent that the cost of repair or restoration exceeds 50 percent of the replacement cost of the sign as determined by the code official.
3. Signs that comply with either Item 1 or 2 need not be permitted.

SECTION 1005 EXEMPT SIGNS

1005.1 Exempt signs. The following signs shall be exempt from the provisions of this chapter. Signs shall not be exempt from Section 1004.4.

1. Official notices authorized by a *court*, public body or public safety official.
2. Directional, warning or information signs authorized by federal, state or municipal governments.
3. Memorial plaques, building identification signs and building cornerstones where cut or carved into a masonry surface or where made of noncombustible material and made an integral part of the building or structure.
4. The flag of a government or noncommercial institution, such as a school.
5. Religious symbols and seasonal decorations within the appropriate public holiday season.
6. Works of fine art displayed in conjunction with a commercial enterprise where the enterprise does not receive direct commercial gain.
7. Street address signs and combination nameplate and street address signs that contain no advertising copy and that do not exceed 6 square feet (0.56 m^2) in area.

SECTION 1006 PROHIBITED SIGNS

1006.1 Prohibited signs. The following devices and locations shall be specifically prohibited:

1. Signs located in such a manner as to obstruct or otherwise interfere with an official traffic sign, signal or device, or obstruct or interfere with a driver's view of approaching, merging or intersecting traffic.
2. Except as provided for elsewhere in this code, signs encroaching on or overhanging public right-of-way. *Signs* shall not be attached to any utility pole, light standard, street tree or any other public facility located within the public right-of-way.
3. Signs that blink, flash or are animated by lighting in any fashion that would cause such signs to have the appearance of traffic safety signs and lights, or municipal vehicle warnings from a distance.
4. Portable signs except as allowed for temporary signs.
5. Any *sign* attached to, or placed on, a vehicle or trailer parked on public or private property, except for signs meeting the following conditions:
    - 5.1. The primary purpose of such a vehicle or trailer is not the display of signs.
    - 5.2. The signs are magnetic, decals or painted on an integral part of the vehicle or equipment as originally designed by the manufacturer, and do not break the silhouette of the vehicle.
    - 5.3. The vehicle or trailer is in operating condition, currently registered and licensed to operate on public streets where applicable, and actively used or available for use in the daily function of the business to which such signs relate.
6. Vehicles and trailers are not used primarily as static displays, advertising a product or service, nor utilized as storage, shelter or distribution points for commercial products or services for the general public.
7. Balloons, streamers or pinwheels except those temporarily displayed as part of a special sale, promotion or community event. For the purposes of this subsection, "temporarily" means not more than 20 days in any calendar year.

SECTION 1007 PERMITS

1007.1 Permits required. Unless specifically exempted, a permit must be obtained from the code official for the erection and maintenance of all signs erected or maintained within this jurisdiction and in accordance with other ordinances of this jurisdiction. Exemptions from the necessity of securing a permit, however, shall not be construed to relieve the owner of the sign involved from responsibility for its erection and maintenance in a safe manner and in a manner in accordance with all the other provisions of this ordinance.

1007.2 Construction documents. Before any permit is granted for the erection of a sign or sign structure requiring such permit, construction documents shall be filed with the code official showing the dimensions, materials and required details of construction, including loads, stresses, anchorage and any other pertinent data. The permit application shall be accompanied by the written consent of the owner or lessee of the premises on which the sign is to be erected and by engineering calculations signed and sealed by a registered design professional where required by the *International Building Code*.

1007.3 Changes to signs. Signs shall not be structurally altered, enlarged or relocated except in conformity to the provisions herein, nor until a proper permit, if required, has been secured. The changing or maintenance of movable parts or components of an approved sign that is designed for such changes, or the changing of copy, business names, lettering, sign faces, colors, display and/or graphic matter, or the content of any sign shall not be deemed a structural alteration.

1007.4 Permit fees. Permit fees to erect, alter or relocate a sign shall be in accordance with the fee schedule adopted within this jurisdiction.

SECTION 1008 SPECIFIC SIGN REQUIREMENTS

1008.1 Identification signs. Identification signs shall be in accordance with Sections 1008.1.1 through 1008.1.3.

1008.1.1 Wall signs. Every single-family residence, multiple-family residential complex, commercial or *industrial* building, and every separate nonresidential building in a residential zone may display wall signs per street frontage subject to the limiting standards set forth in Table 1008.1.1(1). For shopping centers, planned *industrial* parks or other multiple-occupancy nonresidential buildings, the building face or wall shall be calculated separately for each separate occupancy, but in no event will the allowed area for any separate occupancy be less than **[JURISDICTION TO INSERT NUMBER]** square feet.

1008.1.2 Free-standing signs. In addition to any allowable wall signs, every single-family residential subdivision, multiple-family residential complex, commercial or *industrial* building, and every separate nonresidential building in a residential zone shall be permitted to display free-standing or combination signs per street frontage subject to the limiting standards set forth in Table 1008.1.2.

TABLE 1008.1.1(1)
IDENTIFICATION SIGN STANDARDS—WALL SIGNS

LAND USE	AGGREGATE AREA (square feet)
Single-family residential	[JURISDICTION TO INSERT NUMBER]
Multiple-family residential	[JURISDICTION TO INSERT NUMBER]
Nonresidential in a residential zone	[JURISDICTION TO INSERT NUMBER]
Commercial and industrial	See Table 1008.1.1(2)

For SI: 1 square foot = 0.0929 m^2.

TABLE 1008.1.1(2)
SIGN AREA

DISTANCE OF SIGN FROM ROAD OR ADJACENT COMMERCIAL OR INDUSTRIAL ZONE	PERCENTAGE OF BUILDING ELEVATION PERMITTED FOR SIGN AREA
0 to 100 feet	[JURISDICTION TO INSERT NUMBER]
101 to 300 feet	[JURISDICTION TO INSERT NUMBER]
Over 301 feet	[JURISDICTION TO INSERT NUMBER]

For SI: 1 foot = 304.8 mm.

1008.1.3 Directional signs. Not more than two directional signs shall be permitted per street entrance to any lot. There shall be no limit to the number of directional signs providing directional information interior to a lot. In residential zones, the maximum area for directional signs shall be **[JURISDICTION TO INSERT NUMBER]** square feet. For all other zones, the maximum area for any directional sign visible from adjacent property or rights-of-way shall be **[JURISDICTION TO INSERT NUMBER]** square feet. Not more than 25 percent of the area of any directional sign shall be permitted to be devoted to business identification or logo, which area shall not be assessed as identification sign area.

1008.2 Temporary signs. Temporary signs shall be in accordance with Sections 1008.2.1 through 1008.2.6.

1008.2.1 Real estate signs. Real estate signs shall be permitted in all zoning districts, subject to the following limitations:

1. Real estate signs located on a single residential lot shall be limited to one sign, not greater than **[JURISDICTION TO INSERT NUMBER]** feet in height and **[JURISDICTION TO INSERT NUMBER]** square feet in area.
2. Real estate signs advertising the sale of lots located within a subdivision shall be limited to one sign per entrance to the subdivision, and each *sign* shall be not greater than **[JURISDICTION TO INSERT NUMBER]** square feet in area nor **[JURISDICTION TO INSERT NUMBER]** feet in height. Signs permitted under this section shall be removed within 10 days after sale of the last original lot.
3. Real estate signs advertising the sale or lease of space within commercial or *industrial* buildings shall be not greater than **[JURISDICTION TO INSERT NUMBER]** square feet in area nor **[JURISDICTION TO**

INSERT NUMBER] feet in height, and shall be limited to one sign per street front.

4. Real estate signs advertising the sale or lease of vacant commercial or *industrial* land shall be limited to one sign per street front, and each sign shall be not greater than **[JURISDICTION TO INSERT NUMBER]** feet in height, and **[JURISDICTION TO INSERT NUMBER]** square feet for property of 10 acres (40 470 m^2) or less, or 100 square feet (9.3 m^2) for property exceeding 10 acres (40 470 m^2).
5. Real estate signs shall be removed not later than 10 days after execution of a lease agreement in the event of a lease, or the closing of the sale in the event of a purchase.

1008.2.2 Development and construction signs. Signs temporarily erected during construction to inform the public of the developer, contractors, architects, engineers, the nature of the project or anticipated completion dates, shall be permitted in all zoning districts, subject to the following limitations:

1. Such signs on a single residential lot shall be limited to one sign, not greater than **[JURISDICTION TO INSERT NUMBER]** feet in height and **[JURISDICTION TO INSERT NUMBER]** square feet in area.
2. Such signs for a residential subdivision or multiple residential lots shall be limited to one sign, at each entrance to the subdivision or on one of the lots to be built on, and shall be not greater than **[JURISDICTION TO INSERT NUMBER]** feet in height and **[JURISDICTION TO INSERT NUMBER]** square feet in area.
3. Such signs for nonresidential uses in residential districts shall be limited to one sign, and shall be not greater than **[JURISDICTION TO INSERT NUMBER]** feet in height and **[JURISDICTION TO INSERT NUMBER]** square feet in area.
4. Such signs for commercial or *industrial* projects shall be limited to one sign per street front, not to exceed **[JURISDICTION TO INSERT NUMBER]** feet in height and **[JURISDICTION TO INSERT NUMBER]** square feet for projects on parcels 5 acres (20 235 m^2) or less in size, and not to exceed **[JURISDICTION TO INSERT NUMBER]** feet in height and **[JURISDICTION TO INSERT NUMBER]** square feet for projects on parcels larger than 5 acres (20 235 m^2).
5. Development and construction signs shall not be displayed until after the issuance of construction permits by the building official, and must be removed not later than 24 hours following issuance of an occupancy permit for any or all portions of the project.

1008.2.3 Special promotion, event and grand opening signs. Signs temporarily displayed to advertise special promotions, events and grand openings shall be permitted for nonresidential uses in a residential district, and for all commercial and *industrial* districts subject to the following limitations:

1. Such signs shall be limited to one sign per street front.
2. Such signs shall be displayed for not more than 30 consecutive days in any 3-month period, and not more than 60 days in any calendar year. The signs shall be erected not more than 5 days prior to the event or grand opening, and shall be removed not more than 1 day after the event or grand opening.
3. The total area of all such signs shall not exceed **[JURISDICTION TO INSERT NUMBER]** square feet in any single-family residential district, **[JURISDICTION TO INSERT NUMBER]** square feet in any multiple-family residential district and **[JURISDICTION TO INSERT NUMBER]** square feet in any commercial or *industrial* district.

1008.2.4 Special event signs in public ways. Signs advertising a special community event shall not be prohibited in or over public rights-of-way, subject to approval by the code official as to the size, location and method of erec-

TABLE 1008.1.2
IDENTIFICATION SIGN STANDARDS—FREE-STANDING SIGNS[a,b,c]

LAND USE	NUMBER OF SIGNS	HEIGHT (feet)	AREA (square feet)	SPACING
Single-family residential	[JURISDICTION TO INSERT NUMBER]	[JURISDICTION TO INSERT NUMBER]	[JURISDICTION TO INSERT NUMBER]	1 per subdivision entrance[a]
Multiple-family residential	[JURISDICTION TO INSERT NUMBER]	[JURISDICTION TO INSERT NUMBER]	[JURISDICTION TO INSERT NUMBER]	1 per driveway[a]
Nonresidential in a residential zone	[JURISDICTION TO INSERT NUMBER]	[JURISDICTION TO INSERT NUMBER]	[JURISDICTION TO INSERT NUMBER]	300[a]
Commercial and industrial	[JURISDICTION TO INSERT NUMBER]	See Figures 1008.1.2 (1), (2) and (3)	See Figures 1008.1.2 (1), (2) and (3)	150[b]

For SI: 1 foot = 304.8 mm, 1 square foot = 0.0929 m^2, 1 acre = 4047 m^2.

a. For subdivision or apartment identification signs placed on a decorative entry wall approved by the code official, two identification signs shall be permitted to be placed at each entrance to the subdivision or apartment complex, one on each side of the driveway or entry drive.

b. For shopping centers or planned industrial parks, two monument-style free-standing signs not exceeding 50 percent each of the permitted height and area, and spaced not closer than 100 feet to any other free-standing identification sign, shall be permitted to be allowed in lieu of any free-standing sign otherwise permitted in this table.

c. For any commercial or industrial development complex exceeding 1,000,000 square feet of gross leasable area, or 40 acres in size, such as regional shopping centers, auto malls or planned industrial parks, one free-standing sign per street front shall be permitted to be increased in sign area by up to 50 percent.

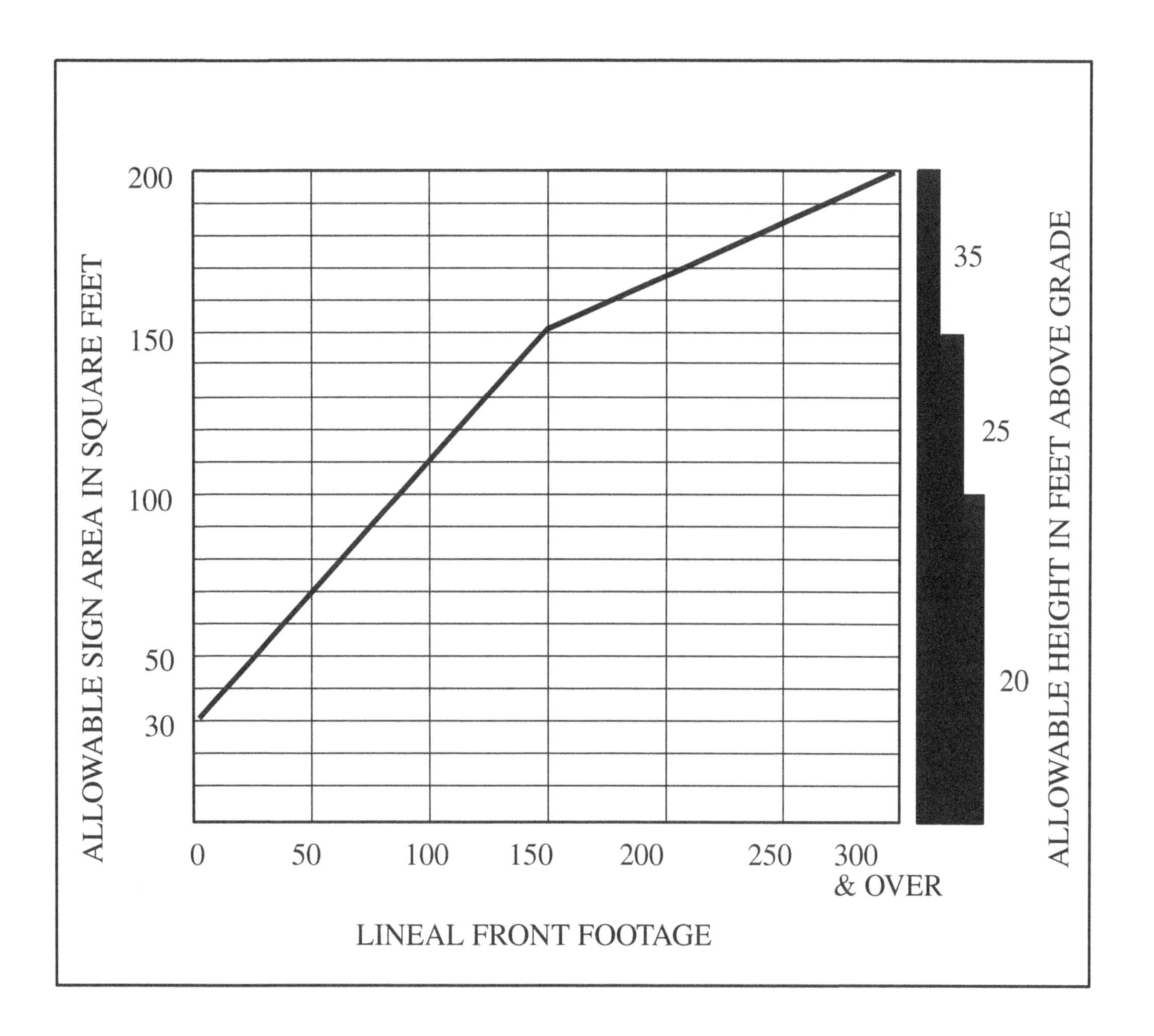

For SI: 1 foot = 304.8 mm, 1 square foot = 0.0929 m^2, 1 mile per hour = 1.609 km/h.

FIGURE 1008.1.2(1)
ON-PREMISE FREE-STANDING SIGNS/COMMERCIAL AND INDUSTRIAL ZONES
VEHICULAR SPEED SUBJECT TO POSTED LIMITS UNDER 35 MILES PER HOUR

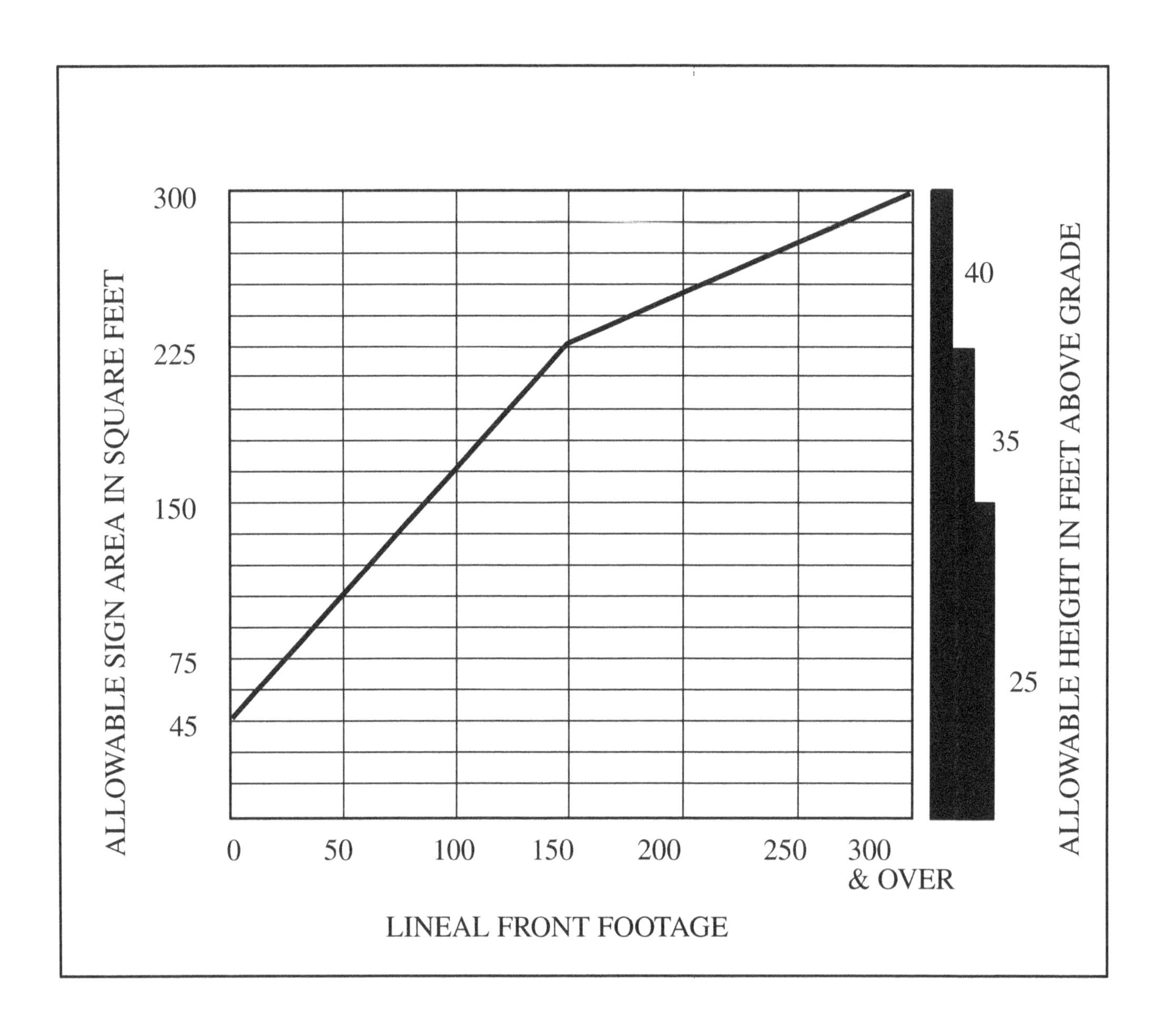

For SI: 1 foot = 304.8 mm, 1 square foot = 0.0929 m^2, 1 mile per hour = 1.609 km/h.

FIGURE 1008.1.2(2)
ON-PREMISE FREE-STANDING SIGNS/COMMERCIAL AND INDUSTRIAL ZONES
VEHICULAR SPEED SUBJECT TO POSTED LIMITS BETWEEN 35 AND 55 MILES PER HOUR (INCLUSIVE)

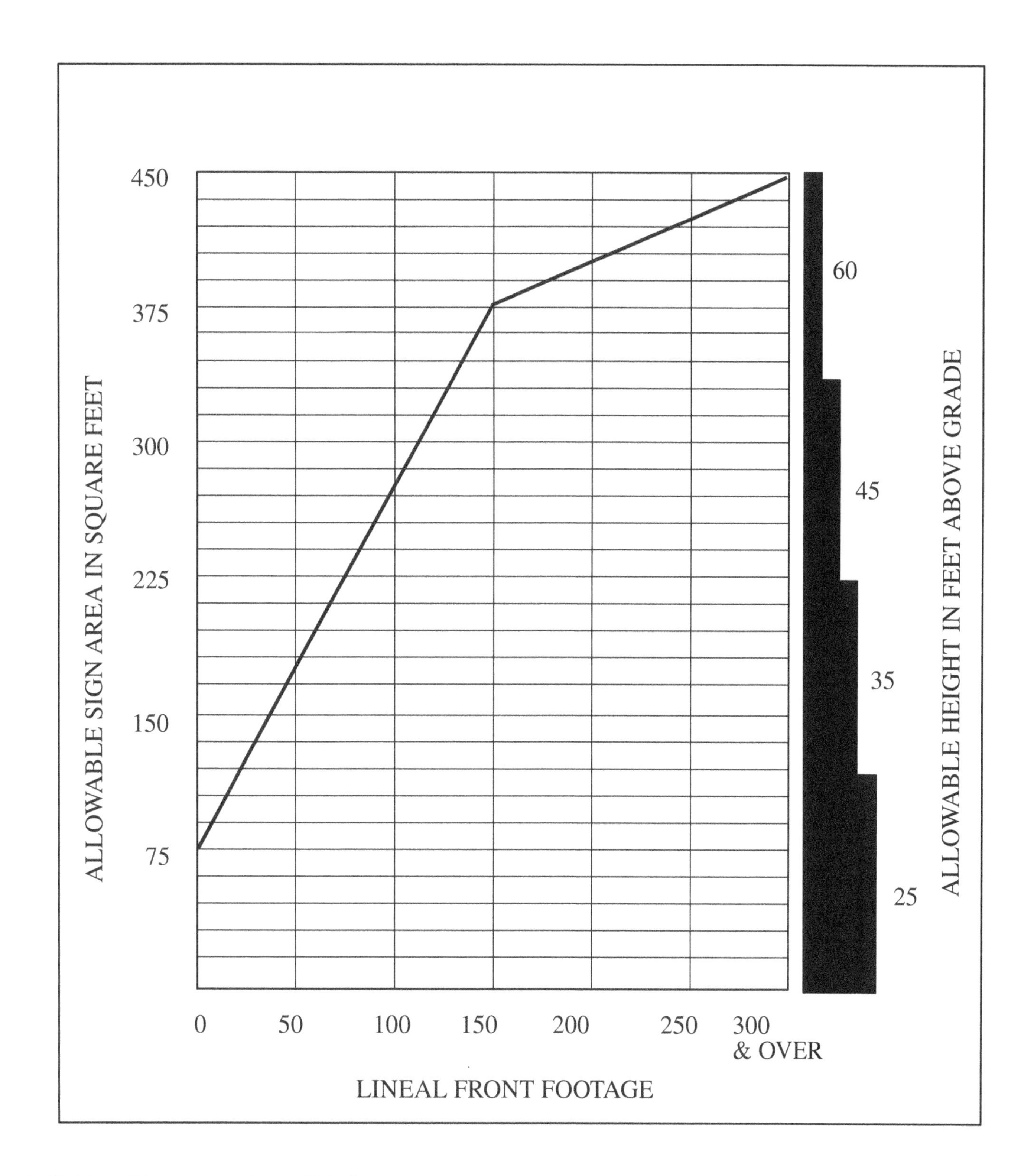

For SI: 1 foot = 304.8 mm, 1 square foot = 0.0929 m^2, 1 mile per hour = 1.609 km/h.

FIGURE 1008.1.2(3)
ON-PREMISE FREE-STANDING SIGNS/COMMERCIAL AND INDUSTRIAL ZONES
VEHICULAR SPEED SUBJECT TO POSTED LIMITS ABOVE 55 MILES PER HOUR

tion. The code official may not approve any special event signage that would impair the safety and convenience of use of public rights-of-way, or obstruct traffic visibility.

1008.2.5 Portable signs. Portable signs shall be permitted only in the C, CR and FI districts, as designated in this code, subject to the following limitations:

1. Not more than one such sign shall be displayed on any property, and shall not exceed a height of [JURISDICTION TO INSERT NUMBER] feet nor an area of [JURISDICTION TO INSERT NUMBER] square feet.
2. Such signs shall be displayed not more than 20 days in any calendar year.
3. Any electrical portable signs shall comply with NFPA 70, as adopted in this jurisdiction.
4. Portable signs shall not be displayed prior to obtaining a sign permit.

1008.2.6 Political signs. Political signs shall be permitted in all zoning districts, subject to the following limitations:

1. Such signs shall not exceed a height of [JURISDICTION TO INSERT NUMBER] feet nor an area of [JURISDICTION TO INSERT NUMBER] square feet.
2. Such signs for election candidates or ballot propositions shall be displayed only for a period of 60 days preceding the election and shall be removed within 10 days after the election, provided that signs promoting successful candidates or ballot propositions in a primary election may remain displayed until not more than 10 days after the general election.
3. Such signs shall not be placed in any public right-of-way or obstruct traffic visibility.

1008.3 Requirements for specific sign types. Signs of specific type shall be in accordance with Sections 1008.3.1 through 1008.3.7.

1008.3.1 Canopy and marquee signs.

1. The permanently-affixed copy area of *canopy* or marquee signs shall not exceed an area equal to 25 percent of the face area of the *canopy*, marquee or architectural projection on which such sign is affixed or applied.
2. Graphic striping, patterns or color bands on the face of a building, *canopy*, marquee or architectural projection shall not be included in the computation of sign copy area.

1008.3.2 Awning signs.

1. The copy area of awning signs shall not exceed an area equal to 25 percent of the background area of the awning or awning surface to which such a sign is affixed or applied, or the permitted area for wall or fascia signs, whichever is less.
2. Neither the background color of an awning, nor any graphic treatment or embellishment thereto such as striping, patterns or valances, shall be included in the computation of sign copy area.

1008.3.3 Projecting signs.

1. Projecting signs shall be permitted in lieu of freestanding signage on any street frontage limited to one sign per occupancy along any street frontage with public entrance to such an occupancy, and shall be limited in height and area to [JURISDICTION TO INSERT NUMBER] square feet per each [JURISDICTION TO INSERT NUMBER] lineal feet of building frontage, except that no such sign shall exceed an area of [JURISDICTION TO INSERT NUMBER] square feet.
2. Such *sign* shall not extend vertically above the highest point of the building facade on which it is mounted by more than [JURISDICTION TO INSERT NUMBER] percent of the height of the building facade.
3. Such signs shall not extend over a public sidewalk in excess of [JURISDICTION TO INSERT NUMBER] percent of the width of the sidewalk.
4. Such signs shall maintain a clear vertical distance above any public sidewalk of not less than [JURISDICTION TO INSERT NUMBER] feet.

1008.3.4 Under *canopy* signs.

1. Under *canopy* signs shall be limited to not more than one such sign per public entrance to any occupancy, and shall be limited to an area not to exceed [JURISDICTION TO INSERT NUMBER] square feet.
2. Such signs shall maintain a clear vertical distance above any sidewalk or pedestrian way of not less than [JURISDICTION TO INSERT NUMBER] feet.

1008.3.5 Roof signs.

1. Roof signs shall be permitted in commercial and *industrial* districts only.
2. Such signs shall be limited to a height above the roofline of the elevation parallel to the sign face of not more than [JURISDICTION TO INSERT NUMBER] percent of the height of the roofline in commercial districts, and [JURISDICTION TO INSERT NUMBER] percent of the height of the roofline in *industrial* districts.
3. The sign area for roof signs shall be assessed against the aggregate permitted area for wall signs on the elevation of the building most closely parallel to the face of the sign.

1008.3.6 Window signs. Window signs shall be permitted for any nonresidential use in a residential district, and for all commercial and *industrial* districts, subject to the following limitations:

1. The aggregate area of all such signs shall not exceed 25 percent of the window area on which such signs are displayed. Window panels separated by muntins or mullions shall be considered to be one continuous window area.
2. Window signs shall not be assessed against the sign area permitted for other sign types.

1008.3.7 Menu boards. Menu board signs shall not be permitted to exceed 50 square feet (4.6 m^2).

SECTION 1009
SIGNS FOR DEVELOPMENT COMPLEXES

1009.1 Master sign plan required. Landlord or single-owner controlled multiple-occupancy development complexes on parcels exceeding 8 acres (32 376 m^2) in size, such as shopping centers or planned *industrial* parks, shall submit to the code official a master sign plan prior to issuance of new sign permits. The master sign plan shall establish standards and criteria for all signs in the complex that require permits, and shall address, at a minimum, the following:

1. Proposed sign locations.
2. Materials.
3. Type of illumination.
4. Design of free-standing sign structures.
5. Size.
6. Quantity.
7. Uniform standards for nonbusiness signage, including directional and informational signs.

1009.2 Development complex sign. In addition to the free-standing business identification signs otherwise allowed by this ordinance, every multiple-occupancy development complex shall be entitled to one free-standing sign per street front, at the maximum size permitted for business identification free-standing signs, to identify the development complex. Business identification shall not be permitted on a development complex sign. Any free-standing sign otherwise permitted under this ordinance may identify the name of the development complex.

1009.3 Compliance with master sign plan. Applications for sign permits for signage within a multiple-occupancy development complex shall comply with the master sign plan.

1009.4 Amendments. Any amendments to a master sign plan must be signed and approved by the owner(s) within the development complex before such amendment will become effective.

CHAPTER 11

NONCONFORMING STRUCTURES AND USES

User note:

About this chapter: *Chapter 11 contains provisions for nonconforming structures and uses regulated under this code. The primary purpose of this chapter is to ensure that existing structures and current land use practices legally established prior to the adoption of this code are allowed to be continued.*

SECTION 1101
GENERAL

1101.1 Continuance. Except as otherwise required by law, a structure or use legally established prior to the adoption date of this code be maintained unchanged. In other than criminal proceedings, the owner, occupant or user shall have the burden to show that the structure, lot or use was lawfully established.

SECTION 1102
DISCONTINUANCE

1102.1 Vacancy. Any lot or structure, or portion thereof, occupied by a nonconforming use, that is or hereafter becomes vacant and remains unoccupied by a nonconforming use for a period of 6 months shall not thereafter be occupied, except by a use that conforms to this code.

1102.2 Damage. If any nonconforming structure or use is, by any cause, damaged to the extent of 50 percent of its value as determined by the code official, it shall not thereafter be reconstructed as such.

SECTION 1103
ENLARGEMENTS AND MODIFICATIONS

1103.1 Maintenance and repair. Maintenance, repairs and structural alterations shall be permitted to be made to nonconforming structures or to a building housing a nonconforming use with valid permits.

1103.2 Changes of nonconforming use. A change of use of a nonconforming use of a structure or parcel of land shall not be made except to that of a conforming use. Where such change is made, the use shall not thereafter be changed back to a nonconforming use.

1103.3 Additions. Additions to nonconforming structures and parking areas shall conform to the requirements of this code. Additions to structures housing nonconforming uses that increase the area of a nonconforming use shall not be made.

CHAPTER 12

CONDITIONAL USES

User note:

About this chapter: *The intent of Chapter 12 is to allow for the occasional need for a use not normally permitted in a particular zoning district and due to the unique characteristics and service that use provides to the public. This chapter establishes the requirements for conditional uses, such as minimum documentation required to support a conditional-use property, conditional-use permits and fees, and the criteria for expiration and revocation of conditional-use permits.*

SECTION 1201 GENERAL

1201.1 Conditional-use permit. A *conditional-use* permit shall be obtained for certain uses, which would become harmonious or compatible with neighboring uses through the application and maintenance of qualifying conditions and located in specific locations within a zone, but shall not be allowed under the general conditions of the zone as stated in this code.

SECTION 1202 APPLICATIONS

1202.1 Submittal. *Conditional-use* permit applications shall be submitted to the code official as provided in this code. Applications shall be accompanied by maps, drawings, statements or other documents in accordance with the provisions of Section 103.7.4. An appropriate fee shall be collected at the time of submittal as determined by the jurisdiction.

SECTION 1203 PUBLIC HEARING

1203.1 Hearing and action. Prior to the approval, amending or denial of a *conditional-use* permit, a public hearing shall be held in accordance with the provisions of Section 109.1. Upon the completion of said public hearing, the commission or examiner shall render a decision within a time limit as required by law.

SECTION 1204 DETERMINATION

1204.1 Authorization. The hearing examiner on appeal, shall have the authority to impose conditions and safeguards as deemed necessary to protect and enhance the health, safety and welfare of the surrounding area. The authorization of a *conditional-use* permit shall not be made unless the evidence presented is such to establish:

1. That such use will not, under the specific circumstances of the particular case, be detrimental to the health, safety or general welfare of the surrounding area and that the proposed use is necessary or desirable and provides a service or facility that contributes to the general well being of the surrounding area.
2. That such use will comply with the regulations and conditions specified in this code for such use.
3. The planning commission or hearing examiner shall itemize, describe or justify, then have recorded and filed in writing, the conditions imposed on the use.

SECTION 1205 EXPIRATION AND REVOCATION

1205.1 General. A *conditional-use* permit shall be considered to be exercised when the use has been established or when a building permit has been issued and substantial construction accomplished. When such permit is abandoned or discontinued for a period of 1 year, it shall not be reestablished, unless authorized by the planning commission, hearing examiner or legislative body on appeal.

A *conditional-use* permit shall be revoked where the applicant fails to comply with conditions imposed by the hearing examiner.

SECTION 1206 AMENDMENTS

1206.1 General. An amendment to an approved *conditional-use* permit shall be submitted to the code official accompanied by supporting information. The planning commission or hearing examiner shall review the amendment and shall be permitted to grant, deny or amend such amendment and impose conditions deemed necessary.

SECTION 1207 CONDITIONAL USE REVIEW CRITERIA

1207.1 General. A request for a *conditional use* shall be approved, approved with conditions or denied. Each request for a *conditional use* approval shall be consistent with the criteria listed in Items 1 through 9 as follows:

1. The request is consistent with all applicable provisions of the comprehensive plan.
2. The request shall not adversely affect adjacent properties.
3. The request is compatible with the existing or allowable uses of adjacent properties.

4. The request can demonstrate that adequate public facilities, including roads, drainage, potable water, sanitary sewer, and police and fire protection exist or will exist to serve the requested use at the time such facilities are needed.
5. The request can demonstrate adequate provision for maintenance of the use and associated structures.
6. The request has minimized, to the degree possible, adverse effects on the natural environment.
7. The request will not create undue traffic congestion.
8. The request will not adversely affect the public health, safety or welfare.
9. The request conforms to all applicable provisions of this code.

CHAPTER 13
PLANNED UNIT DEVELOPMENT

User note:

About this chapter: *Chapter 13 identifies the code requirements for planned unit developments and describes the role of the planning commission. The primary purpose of this chapter is to allow for diversification, variation and imagination in the relationship of uses, structures, open spaces and heights of structures in order to encourage more rational and economic development with relationship to public services, and to encourage and facilitate the preservation of open lands.*

SECTION 1301 GENERAL

1301.1 Approval. *Planned unit developments* (PUDs) shall be allowed by planning commission approval in any zoning district. Such planned unit development permit shall not be granted unless such development will meet the use limitations of the zoning district in which it is located and meet the *density* and other limitations of such districts, except as such requirements may be lawfully modified as provided by this code. Compliance with the regulations of this code in no way excuses the developer from the applicable requirements of a subdivision ordinance, except as modifications thereof are specifically authorized in the approval of the application for the planned unit development.

1301.2 Intent. These regulations are to encourage and provide means for effecting desirable and quality development by permitting greater flexibility and design freedom than that permitted under the basic district regulations, and to accomplish a well-balanced, aesthetically satisfying city and economically desirable development of building sites within a PUD. These regulations are established to permit latitude in the development of the building site if such development is found to be in accordance with the purpose, spirit and intent of this ordinance and is found not to be hazardous, harmful, offensive or otherwise adverse to the environment, property values or the character of the neighborhood or the health, safety and welfare of the community. It is intended to permit and encourage diversification, variation and imagination in the relationship of uses, structures, open spaces and heights of structures for developments conceived and implemented as comprehensive and cohesive unified projects. It is further intended to encourage more rational and economic development with relationship to public services, and to encourage and facilitate the preservation of open lands.

SECTION 1302 CONDITIONS

1302.1 Area. *Planned unit development* shall not have an area less than that approved by the planning commission as adequate for the proposed development.

1302.2 Uses. A *planned unit development* that will contain uses not permitted in the zoning district in which it is to be located will require a change of zoning district and shall be accompanied by an application for a zoning amendment, except that any residential use shall be considered to be a permitted use in a *planned unit development*, which allows residential uses and shall be governed by *density*, design and other requirements of the *planned unit development* permit.

Where a site is situated in more than one use district, the permitted uses applicable to such property in one district may be extended into the adjacent use district.

1302.3 Ownership. The development shall be in single or corporate ownership at the time of application, or the subject of an application filed jointly by all owners of the property.

1302.4 Design. The planning commission shall require such arrangements of structures and open spaces within the site development plan as necessary to ensure that adjacent properties will not be adversely affected.

1302.4.1 Density. *Density* of land use shall in no case be more than 15 percent higher than allowed in the zoning district.

1302.4.2 Arrangement. Where feasible, the least height and *density* of buildings and uses shall be arranged around the boundaries of the development.

1302.4.3 Specific regulations. Lot area, width, yard, height, *density* and coverage regulations shall be determined by approval of the site development plan.

1302.5 Open spaces. Preservation, maintenance and ownership of required open spaces within the development shall be accomplished by either:

1. Dedication of the land as a public park or parkway system; or
2. Creating a permanent, open space *easement* on and over the said private open spaces to guarantee that the open space remain perpetually in recreational use, with ownership and maintenance being the responsibility of an owners' association established with articles of association and bylaws, which are satisfactory to the legislative body.

1302.6 Landscaping. *Landscaping*, fencing and screening related to the uses within the site and as a means of integrating the proposed development into its surroundings shall be planned and presented to the planning commission for approval, together with other required plans for the development. A planting plan showing proposed tree and shrubbery plantings shall be prepared for the entire site to be developed. A grading and drainage plan shall be submitted to the planning commission with the application.

1302.7 Signs. The size, location, design and nature of signs, if any, and the intensity and direction of area or floodlighting shall be detailed in the application.

1302.8 Desirability. The proposed use of the particular location shall be shown as necessary or desirable, to provide a service or facility that will contribute to the general well-being of the surrounding area. It shall also be shown that under the circumstances of the particular case, the proposed use will not be detrimental to the health, safety or general welfare of persons residing in the vicinity of the planned unit development.

SECTION 1303 PLANNING COMMISSION DETERMINATION

1303.1 Considerations. In carrying out the intent of this section, the planning commission shall consider the following principles:

1. It is the intent of this section that site and building plans for a PUD shall be prepared by a designer or team of designers having professional competence in urban planning as proposed in the application. The commission shall be permitted to require the applicant to engage such professional expertise as a qualified designer or design team.
2. It is not the intent of this section that control of the design of a PUD by the planning commission be so rigidly exercised that individual initiative be stifled and substantial additional expense incurred; rather, it is the intent of this section that the control exercised be the minimum necessary to achieve the purpose of this section.
3. The planning commission shall be authorized to approve or disapprove an application for a PUD.

In an approval, the commission shall be permitted to attach such conditions as it deems necessary to secure compliance with the purposes set forth in this chapter. The denial of an application for a PUD by the planning commission shall be permitted to be appealed to the legislative body of the jurisdiction.

SECTION 1304 REQUIRED CONTRIBUTIONS

1304.1 General. The legislative body, as part of the approval of a PUD, shall be permitted to require an applicant to make reasonable contributions to include, but not limited to any combination of the following:

1. Dedication of land for public park purposes.
2. Dedication of land for public school purposes.
3. Dedication of land for public road right-of-way purposes.
4. Construction of, or addition to, roads serving the proposed project where such construction or addition is reasonably related to the traffic to be generated.
5. Installation of required traffic safety devices.
6. Preservation of areas containing significant natural, environmental, historic, archeological or similar resources.

SECTION 1305 PLANNING COMMISSION ACTION

1305.1 Approval. The planning commission shall have the authority to require that the following conditions for a planned unit development (among others it deems appropriate) be met by the applicant:

1. That the proponents intend to start construction within 1 year of either the approval of the project or of any necessary zoning district change, and intend to complete said construction, or approved stages thereof, within 4 years from the date construction begins.
2. That the development is planned as one complex land use rather than as an aggregation of individual and unrelated buildings and uses.

1305.2 Limitations on application.

1. Upon approval of a PUD, construction shall proceed only in accordance with the plans and specifications approved by the planning commission and in compliance with any conditions attached by the jurisdiction as to its approval.
2. Amendment to approved plans and specifications for a PUD shall be obtained only by following the procedures here outlined for first approval.
3. The code official shall not issue any permit for any proposed building, structure or use within the project unless such building, structure or use is in accordance with the approved development plan and with any conditions imposed in conjunction with its approval.

CHAPTER 14
REFERENCED STANDARDS

User note:

About this chapter: *This code contains numerous references to standards that are used to provide requirements for materials and methods of construction. Chapter 14 contains a comprehensive list of all standards that are referenced in this code. These standards, in essence, are part of this code to the extent of the reference to the standard.*

This chapter lists the standards that are referenced in various sections of this document. The standards are listed herein by the promulgating agency of the standard, the standard identification, the effective date and title, and the section or sections of this document that reference the standard.

ICC

International Code Council, Inc.
500 New Jersey Avenue, NW
6th Floor
Washington, DC 20001

IBC—18: International Building Code®
201.3, 202, 1004.1, 1007.2

ICC A117.1—09: Accessible and Usable Buildings and Facilities
801.2.4, 801.3.1

IMC—18: International Mechanical Code®
201.3

INDEX